Soft Energy Paths:
Toward A Durable
Peace

FRIENDS OF THE EARTH ENERGY PAPERS

ECO, volume 2, bound, the 11 papers published in Washington, DC, in 1972 to interpret the AEC hearings on reactor safety.

ECO, volume 3, unbound, the seven-issue series covering meetings of the American Nuclear Society and the Atomic Industrial Forum in San Francisco, November, 1973.

An Assessment of the Emergency Core Cooling Systems Rulemaking Hearings, by the Union of Concerned Scientists.

Cry Crisis! Rehearsal in Alaska, by Harvey Manning.

The Nuclear Fuel Cycle, by the Union of Concerned Scientists.

World Energy Strategies: Facts, Issues, and Options, by Amory B. Lovins:

Non-Nuclear Futures: The Case for an Ethical Energy Strategy, by Amory B. Lovins and John H. Price.

Soft Energy Paths: Toward A Durable Peace, by Amory B. Lovins.

Soft Energy Paths: Toward A Durable Peace

Amory B. Lovins

Friends of the Earth International
San Francisco • New York • Washington, D.C. • London • Paris
Ballinger Publishing Company • Cambridge, Mass.
A Subsidiary of Harper & Row, Publishers, Inc.

This book is printed on recycled paper.

International Standard Book Number: 0-88410-614-4 (C)
0-88410-615-2 (P)

Library of Congress Catalog Card Number: 77-4349

Printed in the United States of America

Library of Congress Cataloging in Publication Data

Lovins, Amory, 1947–
  Soft energy paths.

  1. Power resources. 2. Energy policy. I. Title.
TJ163.2.L678   333.7   77-4349
ISBN 0-88410-614-4 (C)
ISBN 0-88410-615-2 (P)

Contents

Contents vii

※

List of Figures

List of Tables

Foreword

Whether we are talking of an individual citizen or of a whole community, "cataclysmic wealth" can have disastrous consequences. One thinks of all the sad accounts of the miseries that descend on sudden winners of sweepstakes and lotteries—or even of the later years of Howard Hughes. On a grander scale, the golden flood from the conquered Aztec and Inca empires destroyed the balance and sanity of the Spanish state. However valuable a sustained increase in resources may be, it is nearly always a disaster for the shower of gold to arrive in a cloudburst.

Can the disasters be avoided? The question is urgent, for Western society has just been through such a cloudburst—the "black gold" of oil at fifteen cents per barrel at the wellhead—and must now cope with the consequences which, by most historical analogies, can be disastrous. The timeliness of Mr. Amory Lovins's book lies in its direct confrontation of this crisis, its great value in the calm good sense and solid practical value of the alternatives he offers to catastrophe.

There is no secret about the nature of the temptations offered by sudden vast wealth. Its use rises sharply to create new habits and expectations. These habits are accompanied by an irrational lack of care about usefulness or waste. The process develops habits in individual people, and institutions in whole societies, which accustom them to operating on the basis of excess and wastefulness; and, although different episodes have different endings, one prospect sees the affected groups, long after the cloudburst of wealth has passed, trying every kind of expedient—borrowing, sponging, speculating—to try to ensure that the private habits or public institutions of excess and waste are maintained. The result is at best a measure of social disintegration; at worst, collapse.

In the past twenty years, industrialized societies have in varying degrees displayed most of these symptoms. In the 1960s, energy consumption in general grew on average by 5 to 6 percent each year—in Japan by a remarkable 12 percent—with oil's share in commercial energy supplies growing from 39 percent in 1960 to 51 percent in 1974. Owing in part to steadily declining real costs, the growth of demand for electricity was even more rapid—over 7 percent a year between 1958 and 1972—and this in spite of the fact that around three-quarters of the energy used to build and run most power stations is lost in producing and distributing the electricity, a loss that is often not recouped by correspondingly efficient use of the electricity. Private cars, running on cheap gasoline, have risen in number from eighty-nine million in 1960 to around 214 million in 1975. They in turn have displaced less energy-intensive forms of public transport, spread urban settlements thinly over the countryside, and thus imposed the need for further cars.

These new urban patterns have been one factor in creating a rising demand for energy in food production. They entail a very wide network for moving, packaging, freezing, distributing, and finally selling, storing, and cooking food. As a result, most of the energy used is expended after the food leaves the farm. In the U.S. in 1970, only a quarter of the energy used in the food system was consumed on the farm. Yet high mechanization and the overuse of artificial fertilizers, pesticides, and herbicides have ensured that the farms, too, steadily increase their energy demand—not least because urbanization itself tends to displace farming onto less fertile land.

This mixed pattern of rising use and rising waste has had its customary effect. Both people and societies tend to think it normal and believe that the trend should continue. That a nation's wealth and its energy use are virtually two ways of saying the same thing is an example of the kind of folk myth created by sudden affluence. The fact that North Americans use nearly twice as much energy as the not exactly impoverished West Germans—and they, in turn, about twice as much as the far from destitute New Zealanders—is insufficiently known to weaken the fable. Thus all the elements of the customary response to sudden affluence are present—excess, waste, determination to continue, deep belief that the only acceptable way is more of the same and even more of more.

Then comes the crisis. Cheap, accessible supplies of oil and gas are running down. Most sources of new supply—North Sea oil, the Alaskan North Slope—are inherently more expensive to exploit, expensive not only in money but in other and deeper ways. There is a desperate search for means of continuing the old abundance. Although obvious possibilities such as coal have much larger reserves, the means of transforming them into the needed liquid or gaseous fuels are once again comparatively expensive and unpleasant. Yet an economy in which the all-pervasive energy patterns of recent affluence can only be achieved at much higher prices suffers precisely the kind of pressure that sends the pursuers of vanishing wealth off in ever more extreme directions. And there, as they look, stand the temptations of the nuclear economy in which plutonium in fast breeders or, later, tritium and deuterium in fusion reactors will allow—such is the belief—the "Great Energy Jag" to continue.

Like addicts in general, they wave aside the risks. Fusion may still be technically doubtful and carries with it alarming implications of centralized power. It is also not free from problems of radioactive wastes. The even more lethal fission wastes that must follow from a general expansion of uranium-, then plutonium-fueled electricity generation are cumulative since their decay to relatively safe levels takes thousands upon thousands of years and all the time new supplies will be coming into the disposal systems. As the use of plutonium and the stores of waste increase, so does the possibility that, by theft or terrorism, society will be blackmailed and attacked. Behind the whole nuclear venture stands the appalling possibility of nuclear war—possibly the "final solution" itself. No matter. As in the case of alcoholics warned of self-destruction, these possibilities are ignored or denied. The energy addiction is too strong, the alternatives not apparent. Are we destined to repeat the full cycle of cataclysmic wealth, with disaster as the last chapter? Worse still, are we, owing to the deadly nature of the nuclear choice, likely to ensure that our common collapse will be the last?

This is the profound seriousness of the issues that are faced in this book. Mr. Lovins does not in any way underplay the risks. On the contrary, he gives them the stability and certainty of scientific fact. But it is nonetheless a hopeful study, for he shows how pitifully mistaken our energy "alcoholics" are in supposing that no alternatives exist. There are alternatives on every side, and all of them, whether they turn on sane conservation or on the use of safe "energy income," add up to a life quite as convenient as our present existence, certainly more sustainable, and arguably more profoundly satisfying in human terms.

We can cut our waste with such simple measures as returning bottles or building houses that do not leak heat like sieves. We can design city layouts with appropriately sized power stations—as in many Swedish communities—where spare heat from electricity generation warms the houses and heats the water. Industry can provide a much larger share of its own power, and even export a surplus, by "cogeneration" (electricity generated as a byproduct of the process heat normally produced). We can use electricity, with its high cost and potential wastefulness, only for the tasks that genuinely require it—lighting, electronics, electrolysis, motors in industry and home appliances and railways—which together represent less than a tenth of all our end-use energy needs. Heating a house to 68°F with electricity from a nuclear reactor in which fission at trillions of degrees heats the fuel to thousands of degrees is a gratuitous absurdity; and in England, for example, over half our end-use needs are for such low grade purposes as low temperature heat. Much simpler, cheaper, and surer technologies that better match the end uses—space heating, cooking, mobility—are available. Most uses simply do not require the energies needed to produce atomic fireballs. A single nuclear reactor, meticulously engineered, carefully tested, and thoughtfully sited a safe 150 million kilometers away—in fact, the sun itself—is quite enough.

In the place of the hard and unforgiving technologies of nuclear power (coupled with massive and environmentally disruptive growth in fossil fuel consumption), Mr. Lovins proposes a phased, orderly transition to the "soft technologies" based on "energy income." They include using direct solar energy, wind, and biomass conversion —the use of crop, wood, and other organic wastes and, where suitable, perhaps also an ecologically balanced growth of trees and shrubs for conversion to liquid and gaseous fuels. These diverse means can be deployed within half a century to meet the needs of a society that at the same time is using existing systems as a bridge to the future, carrying through all the necessary means of careful resource conservation, suiting its types of energy to the nature of their end use, and learning all over again that the words "thrifty" and "thrive" have a common root.

In short, we do not have to end our recent cataclysmic burst of energy wealth in equally cataclysmic disaster. Mr. Lovins's study, based in the main on American evidence for specificity, proves that the means for a less wasteful, more rational, and more humane future are not only available: they are cheaper and less difficult than the plutonium economy. If other nations were to undertake the same range of review and enquiry, they would be forced to the same conclusions. But they must do their own studies, for the great diversity of energy problems in the world calls for an equal diversity of solutions. In many countries such as India and China, village-scale biogas plants can provide the driving force for rural development. In Brazil, starchy crops such as cassava can and will be used to make vast amounts of fuel alcohols, while in Finland and Québec the same can be done with forestry residues. Wind machines are well suited to Denmark, solar furnaces to Saudi Arabia. Existing large hydroelectric dams can be complemented by new (or resuscitated old) small local hydro sets in Thailand as in New Hampshire. Solar heat in Mexico or Australia needs different technologies than solar heat in Scotland or Sweden, but all are practicable. The concepts behind the integrated food-water-energy systems that work so well in East Asia are being successfully adapted to maritime Canada and will be needed in England.

No two countries, no two parts of the same country, have the same energy needs or opportunities or should seek the same mix of energy technologies. No single book can offer a universal prescription. But it appears, as the exciting search for appropriate answers gains speed, that each country and each person, in different ways, can achieve the same goal: capturing enough of the sun's bounty to give mankind the chance of living—perhaps even happily—ever after.

The conceptual basis for that general conclusion, which national studies can refine and implement in their own diverse ways, is what this book provides. It is the great merit of Mr. Lovins's work that he can convince both the citizen and the scientist that, beyond the great energy outburst, survival is not only possible: it may well be more safe and agreeable as well. This is a conviction that no government, and no voter, can afford to overlook.

—BARBARA WARD

Lodsworth, Sussex
4 March 1977

Acknowledgements

A synthetic paper, like a wall, is mostly bricks. I have added a bit of mortar and art, but the bricks bear many names: notably James Benson, Amasa Bishop, David Brooks, Harvey Brooks, Clark Bullard, Tom Burke, Alex Campbell, Monte Canfield, Jr., Meir Carasso, Hendrik Casimir, Peter Chapman, Tom Cochran, Umberto Columbo, David Comey, Czech Conroy, Charles Correa, Paul Craig, Herman Daly, Ned Dearborn, Peter Dyne, Freeman Dyson, Bent Elbek, Duane Elgin, the late Douglas Elliott, Richard Falk, Bernard Feld, Ian Fells, Jay Forrester, Peter Glaser, Wolf Häfele, Bruce Hannon, Henrik Harboe, Jim Harding, Willis Harman, Denis Hayes, Hazel Henderson, John Holdren, Takao Hoshi, Peter Hunt, Ivan Illich, Barbara Jackson, David Jhirad, William Kellogg, Ann and Henry Kendall, Vladimir Kollontai, Tjalling Koopmans, Lew Kowarski, Gerald Leach, Leon Lindberg, Hans Lohmann, Thomas Long, Måns Lonnroth, Arjun Makhijani, Sicco Mansholt, Bert McInnes, George McRobie, Dennis and Donella Meadows, Niels Meyer, Peter Middleton, Bruce Netschert, Kees Daey Ouwens, Jyoti and Kirit Parikh, Walter Patterson, William Peden, Jerry Plunkett, Alan Poole, Frank Potter, Gareth Price, John Price, William Raup, Amulya Reddy, James Robertson, Gene Rochlin, Rudolph Rometsch, Marc Ross, Charles Ryan, Ignacy Sachs, Lee Schipper, Walter Schroeder, Fritz Schumacher, Robert Socolow, Bent Sørensen, Arthur Squires, Theodore Taylor, John and Nancy Todd, Pierre Trudeau, Eric-Jan Tuininga, William van Arx, Frank von Hippel, Ray Watts, Alvin Weinberg, Jerome Weingart, Andy Wells, Norbert Weyss, Robert Williams, Carroll Wilson, and Frank Zarb. These generous (and often inadvertent) contributors are in no way responsible for the use I have made of their bricks. Any loose, crumbly, or missing bricks have my name on them.

Above all I am grateful to Gerald Barney for suggesting what grew into this book, David Brower for inspiring it, and William Bundy and Jennifer Seymour Whittaker for nurturing Chapter Two far beyond the call of editorial duty. The material which, reshaped, forms the nucleus of Parts Two and Three was commissioned by Dr. Alvin Weinberg and Oak Ridge Associated Universities, neither of whom can be held accountable for my views. ORAU and the Council on Foreign Relations, Inc., have kindly agreed to the recycling of material that they first commissioned.

None of this book could have been written without the generous and tolerant hospitality of Ann and Henry Kendall, of Karl Wendelowski and the staff of the Appalachian Mountain Club's Pinkham Notch Camp, and of Dr. Wolf Häfele and the International Institute for Applied Systems Analysis. I am deeply grateful to them all.

Many of the ideas in this book are exploratory and in transition. They have had the benefit, both before and after they were first published, of extensive review and criticism. I thank particularly all those at Brookhaven, Caltech, the Canadian Government, the US Congress, ERDA, FEA, GAO, Harvard, Harwell, IAEA, IIASA, Los Alamos, Oak Ridge, Petro-Canada, Princeton, and Whiteshell who helped to refine my arguments through seminars and discussions. I must also thank in advance anyone who kindly points out errors of fact or logic that have escaped previous reviews. My permanent forwarding address for this purpose is c/o Friends of the Earth Ltd., 9 Poland Street, London WlV 3DG, England.

—ABL

Laxenburg, Austria
18 February 1977

Part I

Concepts

Chapter 1

Introduction

1.1. TECHNOLOGY IS THE ANSWER! (BUT WHAT WAS THE QUESTION?)

The energy problem, according to conventional wisdom, is how to increase energy supplies (especially domestic supplies) to meet projected demands. The solution to this problem is familiar: ever more remote and fragile places are to be ransacked, at ever greater risk and cost, for increasingly elusive fuels, which are then to be converted to premium forms—electricity and fluids—in ever more costly, complex, centralized, and gigantic plants. The side effects of these efforts become increasingly intolerable even as their output allegedly becomes ever more essential to our way of life and our very survival. As population in most industrial countries rises by less than a fifth over the next few decades, we are told that our use of energy must double and our use of electricity treble. Not fulfilling such prophecies, it is claimed, would mean massive unemployment, economic depression, and freezing in the dark.

But where do these projected "energy needs" come from? Herman Daly provides a pungent but broadly accurate summary:

Recent growth rates of population and per capita energy use [or of population, per capita GNP, and energy use per unit of GNP] are projected up to some arbitrary, round-numbered date. Whatever technologies are required to produce the projected amount are automatically accepted, along with their social implications, and no thought is given to how long the system can last once the projected levels are attained. Trend is,

3

in effect, elevated to destiny, and history either stops or starts afresh on the bi-millenial year, or the year 2050 or whatever.

This approach is unworthy of any organism with a central nervous system, much less a cerebral cortex. To those of us who also have souls it is almost incomprehensible in its inversion of ends and means.

. . . [It says] that there is no such things as enough[;] that growth in population and per capita energy use are either desirable or inevitable[;] that it is useless to worry about the future for more than 20 years, since all reasonably discounted costs and benefits become nil over that period[;] and that the increasing scale of technology is simply time's arrow of progress, and refusal to follow it represents a failure of nerve.[1]

Most thoughtful analysts now see that this approach is rapidly grinding to a halt. It is looking *politically* unworkable: most people, for example, who are on the receiving end of offshore and Arctic oil operations, coal stripping, and the plutonium economy have greeted these enterprises with a comprehensive lack of enthusiasm, because they directly perceive the prohibitive social and environmental costs. Extrapolative policy seems *technically* unworkable: there is mounting evidence that even the richest and most sophisticated countries lack the skills, industrial capacity, and managerial ability to sustain such rapid expansion of untried and unforgiving technologies. And it seems *economically* unworkable: for excellent reasons, such free market mechanisms as still operate have persistently shown themselves unwilling to allocate to the extremely capital-intensive, high risk supply technologies the money needed to build them. The inexorable disintegration of current policy thus makes us reexamine its premises.

The basic tenet of high-energy projections is that the more energy we use, the better off we are. But how much energy we use to accomplish our social goals could instead be considered a measure less of our success than of our failure—just as the amount of traffic we must endure to get where we want to go is a measure not of well-being but rather of our failure to establish a rational settlement pattern. As the U.S. National Research Council CONAES study states: "The first, and dominant, 'facet of the solution' [to the energy problem] relates to the issue of *how fast*, and indeed *whether*, *our use of energy may need to grow and*, ultimately, *how much* energy our society will require to sustain the way of life that it chooses. Energy is but a means to social ends; it is not an end in itself."[2]

[1] H. Daly, "On Thinking About Future Energy Requirements" (Baton Rouge: Department of Economics, Louisiana State University, 1976).
[2] *Interim Report of the National Research Council Committee on Nuclear*

Thus before we conclude that technology—any technology—is the answer, we need to remind ourselves what the question was. In order not to talk nonsense about future energy requirements, we must, as Daly points out,[3] first ask:

1. *Who* is going to require the energy?
2. *How much* energy?
3. *What kind* of energy?
4. For *what purpose*?
5. For *how long*?

Strange though it may seem, such basic questions are just starting to be asked within the energy policy community. Indeed, the first time that professional energy planners ever met in public to discuss scale, centralization, and electrification as policy issues—that is, to question the wisdom of a process of centralized electrification to which hundreds of billions of dollars had already been committed around the world—was in October 1976.[4] But it is the fate of all knowledge, as Huxley said, to begin as heresy and end as superstition. For many analysts, overtaken by events that reversed traditional assumptions, the idea that scale, or electrification, or the details of energy needs are worth discussing has only in the past year emerged from heresy into respectability.

As a result, energy policy today is in ferment. Its atmosphere of intellectual excitement rivals that of the early days of quantum mechanics, with important questions and answers emerging every week and circulating by word of mouth around a lively international grapevine. Much new ground has lately been broken. It is now time to take stock. This book will try to set out some concepts and supporting arguments on which many analysts of diverse views, in many countries, have recently begun to converge.

Of course energy policy is not the only field that has lately been turned upside down. Hallowed assumptions are under challenge everywhere in a world now in rapid transition to an uncertain destination. Even the President of the United States has recently remarked: "We must face the prospect of changing our basic ways of living.

---

and Alternative Energy Systems (Washington, D.C.: National Academy of Sciences, January 1977). (Emphasis in original.)

[3] *Supra* note 1.

[4] This was the Oak Ridge Associated Universities symposium *Future Strategies for Energy Development*, held at Oak Ridge, Tennessee, on 20–21 October 1976. The author's contribution to that meeting provided most of the substance for Parts Two and Three of this book. The proceedings are in press (ORAU, 1977).

This change will either be made on our own initiative in a planned and rational way, or forced on us with chaos and suffering by the inexorable laws of nature."[5] But many who work both on energy policy and in other fields have come to believe that, in this time of change, energy—pervasive, symbolic, strategically central to our way of life—offers perhaps the best integrating principle for the wider shifts of policy and perception that we are groping toward. If we get our energy policy right, many other kinds of policy will tend to fall into place too.

The singular—but inevitable and not unexpected—events of late 1973 that gave energy such prominence in the policy landscape were only the first abrupt steps in an adjustment that will not be finished for decades to come. Ever since the late 1960s, it has been clear that replacing Persian Gulf oil and North American onshore oil and gas in the long run—with nuclear power, coal-based electricity or synthetics, tar sands, oil shale, or even many solar technologies—would cost several times as much as oil cost on the world market. This is still true today. In economic jargon, the long-run marginal cost of energy—that is, how much it will cost to produce an extra oil barrel equivalent (for example) of energy from alternative sources once conventional oil and gas are essentially unavailable —is several times the price of a barrel of OPEC oil today. This has profound implications. Energy, for so long treated as a free good, can no longer be taken for granted,[6] but will become much more expensive no matter what we do. It must now be economized, much as we have economized on costly labor in the recent past. OPEC oil is a bargain, and except for possible short-term fluctuations we shall not have large amounts of energy so cheaply again.

As this realization started to sink in during 1975–1976, we also began to notice in industrial societies some serious structural problems that often seemed especially grave in the energy sector. Centrism, vulnerability, technocracy, repression, alienation, and the stresses and conflicts that they bring now worry many leading politicians. These social and political problems are themselves so important that they seem to many a sufficient reason to seek new approaches to the energy problem.

This book is devoted to a comparison of two energy paths that are distinguished ultimately by their antithetical social implications. To people with a traditional reverence for economics, it might

[5] J.E. Carter, *PE* [Professional Engineer] *Magazine*, December 1976, p. 9.
[6] Nonetheless, many economic models used today to study energy policy come so near to doing this that they do not behave much differently if the price of energy is set equal to zero!

appear that basing energy choices on social criteria is what Kenneth Boulding calls a "heroic decision"—that is, doing something the more expensive way because it is desirable on other and more important grounds than internal economic cost. But surprisingly, a heroic decision does not seem necessary in this case, because the energy system that seems socially more attractive is also cheaper and easier. The technical arguments for this proposition take up much of this book.

1.2. OLD FALLACIES AND NEW INSIGHTS

Developing a new approach to the energy problem and its social ramifications has only become possible as new insights have shown us what questions to ask. Often the questions are so simple that we never thought of asking them before. Consider, for example, the alleged relationship between energy use and prosperity.[7]

Many people who have finally learned that increasing energy efficiency does not mean curtailing functions—that is, that insulating your roof does not mean freezing in the dark[8]—still cling to the bizarre notion that using less energy—or, more often, failing to use much more energy—nevertheless means somehow a loss of prosperity. This idea cannot survive inspection of Table 1-1, which shows how much energy an average person in Denmark used at various times for heating and cooking (over half of all Danish energy end use today). Looking only at the values for 1900 through 1975, one might be tempted to identify increasing energy use with increasing well-being. But if that were true, the statistics for the years 1800 and 1500 would imply that Danes have only just regained the prosperity they enjoyed in the Middle Ages.

Deeper analysis shows what is really happening. In 1500 and 1800 Denmark had a wood and peat economy, and most of the heat went up the chimney rather than into the room or cooking pot—just as it does in the Third World today.[9] In 1900 Denmark

[7] The relationship between primary energy use and GNP is the subject of a large, and largely irrelevant, literature. There are good historical correlations in particular countries, but very large international variations, even between countries in similar circumstances; and anyhow, correlation is not causality. The CONAES economics panel has reportedly concluded that, within very wide limits, energy and GNP need have nothing to do with each other (see the CONAES report [*supra* note 2] due in mid-1977).

[8] Some people still have not learned the difference. For example, Americans for Energy Independence, who advocate "maximum feasible expansion of all forms of energy," are reportedly preparing a conservation program consisting largely of curtailment—presumably to induce the public to identify conservation with the taste of cod-liver oil.

[9] A. Makhijani, "Energy Policy for the Rural Third World" (International

Table 1-1. Average Per Capita Primary Energy Use in
Denmark for Heating and Cooking[a]

Year	Ggcal/y[b]
ca. 1500	7-15
1800	7
1900	3
1950	7
1975	17

[a]Source: S. Bjørnholm et al. (Work Group of the International Federation of Institutes for Advanced Study), "Energy in Denmark 1990-2005: A Case Study," Report 7 (Copenhagen: Niels Bohr Institute, September 1976).
[b]1 Ggcal/y = 133 W.

ran on coal, which was burned efficiently in tight cast-iron stoves, and the useful heat obtained per unit of fuel mined was at its maximum. In 1950 Denmark used mainly oil, incurring refinery losses to run inefficient furnaces. In 1975, the further losses of power stations were added as electrification expanded.

This example shows that a facile identification of primary energy use with well-being telescopes several complex relationships that must be kept separate. How much primary energy we use—the fuel we take out of the ground—does not tell us how much energy is delivered at the point of end use (the device that does the kind of work we desire), for that depends on how efficient our energy supply system is. End-use energy in turn says nothing about how much function we perform with the energy, for that depends on our end-use efficiency. And how much function we perform says nothing about social welfare, which depends on whether the thing we did was worth doing.

A fundamental physical insight we have gained about the energy system, then, is to distinguish primary from end-use energy, and hence to focus on the conversion and distribution losses that rob us of much delivered end-use energy. These losses can be virtually eliminated by determining how much of what kind of energy is needed to do the task for which the energy is desired, and then supplying exactly that kind. This end-use orientation leads to almost the reverse of conventional conclusions about what kinds of energy supply technologies we need to build.

In much the same way, recent research has reversed many widely

Institute for Environment and Development [27 Mortimer St., London W.1, England], September 1976).

held views about energy and the economy: for example, that building many large, costly energy facilities is essential (or at least useful) for reducing national unemployment. If, in the spirit of our five questions about energy requirements, we ask, Whose jobs? What kind of jobs? For how long? then we discover that the people who equated reactors with jobs, and energy thrift with unemployment, had it exactly backwards. In fact, every quintillion joules/year of primary energy fed into new power stations *loses* the U.S. economy some 71,000 net jobs, because power stations produce fewer jobs per dollar, directly and indirectly, than virtually any other major investment in the whole economy.[10] (On this basis each large power station destroys about 4000 net jobs.) It is the conservation, solar, environmental, and related social programs, not the refineries and reactors, that yield the most energy, jobs, and monetary returns per dollar invested. Indeed, energy conservation programs and shifts of investment from energy-wasting to social programs create anywhere from tens of thousands to nearly a million net jobs per quintillion joules/year saved[11]—lasting jobs that use widespread or readily learned skills and need personal initiative and responsibility, not transient jobs requiring exotic skills that are already in short supply.

The huge capital-intensive energy facilities often proposed to relieve unemployment not only make it worse, by draining from the economy the capital that could make more jobs if invested almost anywhere else, but also worsen inflation by tying up billions of dollars nonproductively for a decade. And unemployment and inflation are only the first of a long list of distressing side effects of a high growth, high technology, high risk approach to our energy problems. Recent analyses[12] have started to trace the complex

[10] B. Hannon, "Energy and Labor Demand in the Conserver Society" (Urbana-Champaign: Center for Advanced Computation, University of Illinois, July 1976). See also S. Laitner, "The Impact of Solar and Conservation Technologies Upon Labor Demand," Public Citizen (Box 19404, Washington, D.C. 20036), May 1976; and R.D. Scott, "The Energy Dilemma—What It Means to Jobs," International Woodworkers of America (1622 N. Lombard, Portland, Oregon 97217), 1976.

[11] *Ibid.*

[12] For example, L.M. Lindberg et al., *The Energy Syndrome: Comparing National Responses to the Energy Crisis* (Lexington, Massachusetts: Lexington Books, 1977); J.R. Hammarlund and L.N. Lindberg, eds., "The Political Economy of Energy Policy: A Projection for Capitalist Society" (Madison: Institute for Environmental Studies, University of Wisconsin, IES-70, December 1976), chs. 1 and 4-6; in a brief convergent analysis, B. Fritsch, "The Future of the World Economic Order," Arbeitspapier 76/4, Institut für Wirtschaftsforschung, Eidgenössliche Technische Hochschule Zürich, 1976; K.E.F. Watt et al., "The Long-Term Implications and Constraints of Alternate Energy Policies," January 1976, in Subcommittee on Energy and Power, Commerce Committee, USHR,

connections between many of the everyday problems whose relationship to the energy problem is obvious only once it is pointed out.

Consider, for example, some of the interactive effects of a policy based on rapidly increasing the use of energy while holding its price down through regulation and subsidy. Americans—others can draw their own analogies—might see the consequences thus:

We use the apparently cheap energy wastefully, and thus continue to increase our dependence on imported oil, to the detriment of the Third World, Europe, Japan, and our own independence. We earn the foreign exchange to pay for the oil by running down domestic stocks of commodities, which is inflationary; by exporting weapons, which is inflationary, destabilizing, and immoral; and by exporting wheat and soybeans, which inverts midwestern real estate markets, makes us mine groundwater unsustainably in Kansas, and raises our own food prices. Exported American wheat diverts Soviet investment from agriculture into defense, making us increase our own inflationary defense budget, which we have to raise anyhow to defend the sea lanes to bring in the oil and to defend the Israelis from the arms we sold to the Arabs.

With crop exports crucial to our balance of payments, pressure increases for energy- and water-intensive agribusiness.[13] This creates yet another spiral by impairing free natural life support systems and so requiring costly, energy-intensive technical fixes (such as desalination and artificial fertilizers) that increase the stress on remaining natural systems while starving social investments. Excessive substitution of apparently cheap inanimate energy for people causes structural unemployment, which worsens poverty and inequity, which increase alienation and crime. Priorities in crime control and health care are stalled by the heavy capital demands of building and subsidizing the energy sector, which itself contributes to the unemployment and illness at which these social investments were aimed. The drift toward a garrison state at home (needed to protect the vulnerable energy system from strikes, sabotage, and dissent), failure to address rational development goals abroad, and the strengthening of oligopolies and of oil and uranium cartels all encourage international distrust and domestic dissent, both entailing further suspicion and repression. Energy-related climatic shifts could jeopardize

Hearings 94-63 (Washington, D.C.: USGPO, 1976), pt. 1, March 1976, pp. 305-467 (and earlier testimony in same volume).

[13] R.A. Herendeen and C.W. Bullard III, *En. Systs. & Pol.* *1*:383-90 (1976), find that though the amounts of energy embodied in all goods imported and exported by the U.S. are nearly equal, the latter is 38 percent of net fuel imports. Agribusiness is thus already a significant term in U.S. energy trade. Its energy costs may rise steeply as side effects increase and groundwater is depleted.

marginal agriculture, even in the midwestern breadbasket, endangering an increasingly fragile world peace. The nuclear weapons proliferated by the widespread use of nuclear power, the competitive export of arms, reactors, and inflation from rich to poor countries, and the tensions of an ever more inequitable, hungry, and anarchic world could prove a deadly combination.

If it were true, as the proponents of extrapolative high energy futures insist, that there is no alternative, then the human prospect would be bleak indeed, and there would be no point in energy policy or anything else. But a coherent alternative, indeed a family of alternatives (called "soft energy paths"), does exist. It is the task of this book to explore soft energy paths by outlining, in Chapter Two, two hypothetical and illustrative energy futures for the United States—not as forecasts or precise recommendations, but rather as a qualitative vehicle for ideas relevant to many, even to most, countries.

In order to avoid the danger Niels Bohr warned about—"speaking more clearly than we think"—Chapters Three through Eight will specify technical arguments and documentation for conclusions summarized in Chapter Two. Chapters Nine through Eleven will then offer a semitechnical supplement to the later sections of Chapter Two, and the Afterword will place this book in the context of continuing international research. The entire text and its ample annotation are at a level useful both for interested citizens without a technical background and for professional analysts who want to dig deeper. The few sections in Part Two that are too technical to be useful to the former audience can be skipped or skimmed without losing the thread of the argument. A Glossary at the very end, though nowhere referred to in the text, defines symbols, acronyms, and a few technical terms that lay readers might find unfamiliar.

1.3. A PERSONAL NOTE

Nobody can make a completely value-free analysis. The questions that an analyst asks are conditioned by underlying ideas about energy and social problems—though, to protect the unwary reader from answers that are likewise conditioned, the assumptions, calculations, and documentation should be fully set out, as I try to do in later chapters. I do not pretend here to neutrality: but not for the reason some might suppose. If I seem to be presenting advocacy as well as analysis, it is not because I began with a preconceived attachment to a particular ideology about energy or technology, such as the "small is beautiful" philosophy that some have tried to read into my results. It is instead because the results of the analysis so impressed me. This is not an apology: there is nothing wrong with ideologies so long as one is aware of them, and I do think Schumacher[14] makes sense. But as one brought up in the high technology tradition, I was surprised that the analysis made a much stronger case for much more unconventional conclusions than I had expected.

Underlying much of the energy debate is a tacit, implicit divergence about what the energy problem "really" is. Public discourse suffers because our society has mechanisms only for resolving conflicting interests, not conflicting views of reality,[15] so we seldom notice that those perceptions differ markedly. I see no basis for deciding (in the absence of another century's experience mellowed by 20/20 hindsight) which of the several prevalent world views, if any, is most useful, let alone which is "right" or "wrong." As a basis for mutual understanding, therefore, instead of leaving my world view to be guessed at (as most energy writers do), I shall make explicit a few of my underlying opinions—not on every aspect of the whole universe of perceptions that must support any coherent view of our energy future, but at least on a few basic values. Attempting this is unusual and difficult but important. Briefly, then, I think that:

1. we are more endangered by too much energy too soon than by too little too late, for we understand too little the wise use of power;

2. we know next to nothing about the carefully designed natural systems and cycles on which we depend[16]; we must therefore take care to preserve resilience and flexibility, and to design for large safety margins (whose importance we do not yet understand), recognizing the existence of human fallibility, malice, and irrationality (including our own) and of present trends that erode the earth's carrying capacity;

3. people are more important than goods; hence energy, technol-

[14] E.F. Schumacher, *Small Is Beautiful*, (New York: Harper & Row, 1973). His follow-up, *Small Is Possible*, is in preparation. See also the publications of the organizations listed by TRANET (c/o Ellis, 7410 Vernon Sq. Dr., Alexandria, Virginia 22306) and of the Institute for Local Self-Reliance (1717 Mass. Ave. NW, Washington, D.C.); *Liklik Buk Bilong Kain Kain Samting* (P.R. Hale et al., Box 122, Kavieng, New Ireland, Papua-New Guinea), 1976, probably the best of the appropriate technology "catalogs"; and the many studies noted in the access newsletter RAIN (2270 N. Irving, Portland, Oregon 97210). These references, though scant, may give at least a glimpse of the rich diversity of appropriate technology research.

[15] J.S. Reuyl, W. Harman, et al., *A Preliminary Social and Environmental Assessment of the ERDA Solar Energy Program 1975-2020* (Stanford Research Institute draft submitted to ERDA Solar Division, July 1976).

[16] A.B. Lovins, *Environmental Conservation* (Geneva) *3*:1, 3-14 (1976).

ogy, and economic activity are means, not ends, and their quantity is not a measure of welfare; hence economic rationality is a narrow and often defective test of the wisdom of broad social choices, and economic costs and prices, which depend largely on philosophical conventions (see Chapter 3.2), are neither revealed truth nor a meaningful test of rational or desirable behavior;

4. though the potential for growth in the social, cultural, and spiritual spheres is unlimited, resource-crunching material growth is inherently limited (a consequence of the round-earth theory) and, in countries as affluent as the U.S., should be not merely stabilized but returned to sustainable levels at which the net marginal utility of economic activity (to borrow for a moment the economist's abstractions) is clearly positive;

5. since sustainability is more important than the momentary advantage of any generation or group, long-term discount rates should be zero or even slightly negative, reinforcing a frugal, though not penurious, ethic of husbanding;

6. the energy problem[17] should be not how to expand supplies to meet the postulated extrapolative needs of a dynamic economy, but rather how to accomplish social goals elegantly with a minimum of energy and effort, meanwhile taking care to preserve a social fabric that not only tolerates but encourages diverse values and lifestyles;

7. the technical, economic, and social problems[18,19,20] of fission technology are so intractable, and technical efforts to palliate those problems are politically so dangerous, that we should abandon the technology with due deliberate speed;

8. many other energy technologies[21] are exceedingly unattractive

[17] A.B. Lovins, *World Energy Strategies: Facts, Issues, and Options* (Cambridge, Massachusetts: FOE/Ballinger, 1975).

[18] A.B. Lovins & J.H. Price, *Non-Nuclear Futures: The Case for an Ethical Energy Strategy* (Cambridge, Massachusetts: FOE/Ballinger, 1975).

[19] W.C. Patterson, *Nuclear Power* (Harmondsworth, Middlesex, England: Pelican original, Penguin, 1976).

[20] Royal Commission on Environmental Pollution (Sir Brian Flowers, Chairman), *Sixth Report: Nuclear Power and the Environment*, Cmnd. 6618 (London: HMSO, September 1976); Sir Brian Flowers, "Nuclear Power and Public Policy," speech to British Nuclear Energy Society, 2 December 1976; Ranger Uranium Environmental Inquiry (Mr. Justice R.W. Fox, Presiding Commissioner), *First Report* (Canberra: Australian Government Publishing Service, 1976); J.M. Brown, "Health, Safety, and Social Issues," in *The California Nuclear Initiative* (Palo Alto, California: Stanford University Institute for Energy Studies, April 1976); J. Francis and P. Abrecht, eds., *Facing Up to Nuclear Power* (Edinburgh: St. Andrew Press, and Philadelphia: Westminster Press, 1976).

[21] *Supra* notes 16 and 17.

and should be developed and deployed sparingly or not at all (such as nuclear fusion, large coal-fired power stations and conversion plants, many current coal-mining technologies, urban-sited terminals for liquefied natural gas, much Arctic and offshore petroleum extraction, most "unconventional" hydrocarbons, and many "exotic" large-scale solar technologies such as solar satellites and monocultural biomass plantations);

9. ordinary people are qualified and responsible to make these and other energy choices through the democratic political process, and on the social and ethical issues central to such choices the opinion of any technical expert is entitled to no special weight; for although humanity and human institutions are not perfectable, legitimacy and the nearest we can get to wisdom both flow, as Jefferson believed, from the people, whereas pragmatic Hamiltonian concepts of central governance by a cynical elite are unworthy of the people, increase the likelihood and consequences of major errors, and are ultimately tyrannical;

10. issues of material growth are inseparable from the more important issues of distributional equity, both within and among nations; indeed, high growth in overdeveloped countries is inimical to development in poor countries;

11. for poor countries, the self-reliant ecodevelopment concepts inherent in the New Economic Order approach are commendable and practicable while the patterns of industrial development that served the OECD countries in the different circumstances of the past two centuries are not: indeed, so much have conditions changed that ecodevelopment concepts are now the most appropriate for the rich countries too;

12. national interests lie less in traditional geopolitical balancing acts than in striving to attain a just and equitable, therefore peaceful, world order, even at the expense of temporary commercial advantage.

Perhaps these elements of a credo will illustrate what lies behind my views on energy strategy, just as other perceptions lie behind different views. (Many engineers, for example, would lay more emphasis on human perfectability in the technical sphere and less in the social or moral sphere.) But my views on energy and on other matters coevolve; neither can be said to be derived from, or reshaped to justify, the other.

Nor must arguments that stand on their own in Chapter Two have the above beliefs read into them, thus giving them a meaning they will not bear. For example, despite point 4 above, the improved energy efficiencies described in section 2.4 and in Figure 2-2 can be construed as arising *entirely* from technical changes without significant effect on lifestyles. People who consider, as I do, that today's values and institutions are imperfect are welcome to assume instead some mixture of technical and social changes—perhaps substituting repair and recycling for the throwaway economy, for example. But no such assumption is necessary to the argument in section 2.4. Likewise, since Chapter Two argues that nuclear power is impracticably capital-intensive, unnecessary, and an encumbrance, point 7 above, being lower in the hierarchy of nuclear issues, need in no way enter into the argument.

1.4. ASYMMETRIES IN POLICY

The future is no more certain today than it has ever been, and we hear much about making decisions under uncertainty. But to do this properly we must be wary of the danger of not being imaginative enough to see how undetermined the future is and how far we can shape it. As Kenneth Boulding remarks, deciding under uncertainty is bad enough, but deciding under an illusion of certainty is catastrophic.

It is therefore important not to reject out of hand futures that are clearly possible on the basis of Boulding's First Law ("Anything that exists is possible"). As Daly reminds us,

> . . . the "low quad scenario" (one half current U.S. per capita [primary] energy usage) *exists* today in Western Europe. It also existed in the United States as recently as 1960. Therefore the common notion that the low demand scenario is "far out"[,] or merely a hypothetical polar case, is due to inability to recognize the obvious. It is the high demand, hard technology scenarios that have never before been experienced and are completely hypothetical. Yet our "crackpot realists" all treat the hypothetical high energy projection as if it were empirically verified, and the empirically verified low energy scenario as if it were the flimsiest conjecture!
>
> . . . Many people do not like to face up to this basic choice because it is not a question of rationality of means, but of sanity of ends. Taking a position requires moral self-definition, imposes responsibility, and may involve one in conflict. At this stage of the discussion refusal to take a position, accompanied by the usual call for "more studies," serves only to increase the already excessive output of unconscionable mush.[22]

[22] *Supra* note 1. (Emphasis in original.)

Conversely, it is equally important not to rely on Mildew's Dictum[23] ("There is no precedent for anything until it is done for the first time") unless one does so symmetrically. For example, many an official is happy to base national energy policy on someone's guess that there is a probability x that in the year y someone will find z barrels of oil someplace where no oil has yet been found. Yet if one suggests to the same official that policy package a, demonstrated in countries b and c, will add at least d centimeters of roof insulation to e houses and thus save f joules of heat equivalent to g barrels of oil imports, he will argue vigorously that that is wholly speculative and an inadequate basis for policy! Yet it is the sort of number that one knows if one knows any sort of number.

This asymmetry is pervasive in energy policy today. It is deeply troublesome. It emerges in such simple guises as assuming that increased energy efficiency or roof insulation will alter lifestyles while increased energy use or a plutonium economy will not; or that the technical success of district heating or wind conversion is speculative while that of large fast breeder reactors or coal gasification plants is not. Indeed, one is commonly told that we should plan to rely on proven technologies that are already here and known to work, not on speculative "pie in the sky" technologies that are not here and may or may not work. But which are which?

An especially common version of this fallacious asymmetry arises in the fuel-poor but energy-rich countries of Europe: for example, France, which has little coal and no oil or gas, but ample installed hydroelectricity and abundant sunlight and biomass. From a French perspective, the approach described in Chapter Two may seem irrelevant, because the transitional domestic fuels are not available. The only interesting question, in this view, is how to relieve dependence on imported oil as soon as possible (preferably without substituting dependence on, say, imported uranium—a point France seems to have forgotten). Many French analysts argue from these premises that relieving oil dependence with solar technologies takes a great deal of time and money, and should therefore not be considered. Of course it takes much time and money; but so do nuclear technologies. That is why Chapter Two is highly relevant to countries like France. The absence of French transitional fuels akin to U.S. coal and petroleum does not make soft technologies less relevant to France; after all, whatever conservation and transitional supply can be arranged will be a good idea with or without

[23] Lord Mildew was a mythical judge created by the late Sir Alan Herbert (A.P. Herbert), a prolific English author of spoof cases. This dictum is quoted in his *Uncommon Law* (London: Methuen, 1935/1970).

nuclear or solar energy. Such measures do not wait on the nuclear decision (though nuclear investments may inhibit them). The real question is which can relieve oil dependence faster—nuclear power or the "soft" renewable technologies (see Chapter 2.5). That is, can a franc invested in biomass conversion, say, or in solar heat be relied upon to reduce oil imports faster than a franc invested in reactors (having due regard to the very different side effects of both)? The answer to this complex question appears to be yes, for fundamental reasons discussed later in this book.

Indeed, the first countries analyzed in the style of Chapter Two were Japan and Denmark, both almost completely dependent on imported oil. For both, as mentioned elsewhere,[24] a "soft" energy path with no nuclear power appeared quicker and cheaper than present policy, with quicker relief of oil dependence meanwhile. Later independent analyses of the Danish case[25] have provided more detailed data that, taken together, converge well with my earlier estimates, and support the view that nuclear power, if it can be a route to energy independence at all, is not the quickest route available.

Still more recently, a continuing regionalized study by the Science Policy Research Unit of the University of Sussex has reached the preliminary conclusion that within about seventy-five years, the world can be living comfortably within its energy income, using no nuclear power (and certainly needing no breeder reactors meanwhile). The question SPRU is just starting to explore is what the transitional period should look like and whether it requires a nuclear component. Such analysis is complex and depends greatly on local circumstances. But already it seems hard to argue convincingly that a nuclear-powered transition would be quicker, for countries with no fossil fuels of their own, than the contribution that diverse renewable sources could instead make during the same period at similar or lower cost.

As sections 2.6 and 2.11 of Chapter Two suggest, these two choices exclude each other by logistical, cultural, and institutional competition. Though they are not technically incompatible, they

---

[24] *Supra* note 18.
[25] Work Group of The International Federation of Institutes for Advanced Study, "Energy in Denmark 1990–2005: A Case Study," report 7, c/o Sven Bjørnholm, Niels Bohr Institutet (København), September 1976; S. Blegaa et al., "Skitse til alternativ energiplan for Danmark" (Copenhagen: OOA [Skindergade 26, 1, 1159 København K] / OVE [Arendalsgade 3, Kld., 2100 København Ø], 1976; O. Danielsen et al., *Alternative energikilder og -politik* (Copenhagen: Energi Oplysning Utvanget, Grundbog nr. 6, August 1975); and the original projection of B. Sørensen, *Science 189*:255-60 (1975).

are deeply incompatible in all the other ways that matter, and we cannot do both at once. Thus nuclear power, as Walter Patterson puts it, is not a yes-no question, but an either-or question: Do we have it, or do we instead have the other systems with which it competes for our resources?

This question of how much our resources—including social resources—allow us to do leads to the deeper question of how much we need to do. Few people oppose solar energy in its many forms, nor improved energy efficiency, nor the modest and intelligent use of coal under strict environmental controls. Chapter Two argues that if we combine all these measures in the right way, then we shall simply *not need* most of the big supply technologies debated today—especially the costliest, nastiest, and riskiest ones. It is hard for people who have spent their lives developing such systems to accept the idea that their handiwork is now superfluous; but their devices are clearly too expensive and too dangerous to be suitable toys to gratify even the most deserving technologists.

In short, if we do our sums, we shall find that energy supply people today want to supply more energy than serious students of energy demand are demanding—or can see any rational use for. If we buy an excellent raincoat (as Chapter Two proposes in the form of efficient use of energy from a wide range of alternative sources), then we shall not need to use an umbrella, a roof, and weather modification too[26] (in the form of reactors, synthetic-fuel plants, etc.). Nor can we afford them all. Of course we need diversity, but we cannot do everything, and if we try, practical constraints will force us to choose priorities. Priorities inevitably advance some options while retarding others.[27] Some options even foreclose others. Chapter Two suggests that some *combinations* of options can mesh to form a coherent policy consistent with our needs, while other combinations produce merely a hash.

1.5. HOW CAN WE GET THERE FROM HERE?

Who needs how much of what kind of energy for what purpose for how long? If the view argued in the next ten chapters is correct, addressing this basic question suggests a different path along which our energy system can evolve from now on: a way of redirecting

[26] H.A. Bethe and A.B. Lovins, exchange of letters, *Foreign Affairs* 55: 636–40, April 1977.

[27] A.B. Lovins, testimony to President's Council on Environmental Quality, Hearings on the ERDA RD&D Plan (Washington, D.C.: 3 September 1975).

our efforts, thus disproportionately freeing resources for other tasks that can use them more effectively. To do this requires us to take three initial steps that will enable ordinary market and social processes to complete the job.

The first of these steps is correcting the institutional barriers (see Chapter 2.4 and elsewhere) that now impede conservation and rational supply technologies. The second step is removing the subsidies now given to conventional fuel and power industries—now estimated[28] at well over $10 billion per year in the United States alone[29] —and vigorously enforcing antitrust laws. The third step is gradually making energy prices consistent with what it will cost in the long run to replace our dwindling stocks of cheap fuels (see Chapters 3 and 8).

This last proposal[30],[31] needs amplification. Section 1.2 above suggested that cheap energy is really an illusion for which we pay dearly everywhere else in the economy: that it causes structural distortions and long-range indirect effects that are very expensive. Even if this is not so, however, we are deceiving ourselves if we pretend that we really save money by charging ourselves, say, a few dollars per oil barrel equivalent for natural gas—so that we use it up quickly and wastefully—when we know it will cost over $30 per barrel to deliver a synthetic replacement for it once it has run out. Precisely this self-deception led to the disruptive 1976-1977 winter gas shortages in the U.S.

The energy system is so vast and complex that putting in place any replacement technology for natural gas will take decades.[32] Many countries are already considering how to do this. The U.S. Congress has already been asked for billions of dollars to subsidize the synthesis of gas from coal, because the days of cheap gas are numbered, and "synthetic natural gas," even at over $30 per barrel equivalent, will be a bargain when there is no more natural natural gas to be had. Since we are obliged to start committing money now to long-term replacement technologies for the coming decades, we should compare the cost of those technologies with the cost of

---

[28] *Supra* note 15.

[29] C. and J. Steinhart, *Energy: Sources, Use and Role in Human Affairs* (North Scituate, Massachusetts: Duxbury Press, 1974), estimate that about a sixth of an average U.S. family's energy bill is paid compulsorily through taxes used to subsidize the energy sector.

[30] D.W. Orr et al., "The Wolfcreek Statement: Toward a Sustainable Energy Society" (Atlanta: Georgia Conservancy, January 1977).

[31] Energy Policy Project of the Ford Foundation, *A Time to Choose: America's Energy Future* (Cambridge, Massachusetts: Ballinger for The Energy Policy Project, 1974).

[32] *Supra* notes 17 and 18.

other long-run methods of doing the same task, such as the solar heating and thermal insulation that we could use to heat buildings instead. Such a comparison (see Chapters 2, 3, and 8) is relevant if we want to use our resources wisely. Whether solar heat can compete today with a rapidly vanishing supply of artificially cheap gas is irrelevant and misleading—even though it is the kind of comparison on which virtually every study of solar economics and solar "market penetration rates" is based.

Suppose, as an example in round numbers, that we are charging ourselves $10 per barrel for oil that will cost $30 per barrel to replace, and that the replacement technology will be needed soon enough on such a scale that we must start building it now. Today, however, the replacement can only be sold at a $20 per barrel loss. There are two ways out. One is to subsidize the replacement at $20 per barrel by a wide variety of overt or covert methods; but this encourages wasteful use by making us think the oil is cheaper than it really is (as many nations have done for years with natural gas), and it also adds to the burden on taxpayers, who may not be the same people who use most of the oil. A more rational method would be to start now—gradually, but in such a way that everyone knew in advance the schedule of increase—to charge ourselves more, ul- timately $30 per barrel, for all oil, both present and replacement. That would encourage efficient use; would keep in circulation within our own economy $20 per barrel that might otherwise worsen the balance of payments deficit; and would let us smoothly antici- pate the inevitable price rise rather than having to swallow it all at once later when we are less well prepared for it. Gradualism does not dilute the effect of the tax, since price matters less than people's perception of how price will behave. The price need only rise faster than wages and interest rates.

A suitable method of charging ourselves the extra $20—equitably and without difficult administration—would be to phase in a federal tax, which economists might call a "severance royalty" and engineers a "BTU tax." (The latter name is less suitable: BTUs are obsolete, and the name implies a tax on energy generally, not on depletable fuels only.) It would be charged on all fuels, according to their energy content, as they came out of the ground or into the coun- try. It would, in essence, be a depletion allowance backwards. Its effects would work through the economy and be reflected auto- matically in the price of goods and services in proportion to their direct and indirect energy content.

Such a tax would not in itself be redistributive, since, at least in the U.S., the fraction of income spent directly and indirectly on energy does not vary significantly with income.[33] But the resulting revenues could be used redistributively, for example to finance a minimum income or negative income tax program, as well as to encourage energy conservation. Such uses could be politically attractive—especially if combined with general tax and welfare reform and with elimination of the energy subsidies formerly paid through taxes.[34]

We know a good deal about how such a tax might work in the U.S., not only because it would pass through the region of energy prices common in Europe today, but because it would return the U.S. to the kinds of energy prices (in terms of constant purchasing power) that Americans paid only a few decades ago and are paying today for some fuels (e.g., gasoline is now priced in the U.S. at about $30 per barrel, and in Europe at $50-80 per barrel). Cogent arguments can be made[35] that an intelligently arranged fuel tax would help to cope with inflation and unemployment. Of course, some—not all—of the structural benefits of an energy tax in correcting past distortions throughout the economy might be obtained, less elegantly and safely, by deregulating prices. This is not the place to reach into that notorious can of peculiarly North American worms. It is worth remarking, though, that for deregulation to benefit the public rather than merely the "oiligopolists," one must legislate and enforce a truly ratproof windfall profits tax, and so far nobody has figured out how to do this.

Let us suppose that, as is the case now, we sell ourselves energy at unrealistically low prices uncorrected by a proper tax. We are then faced with a quandary. Consider our earlier example, but suppose that we have the option not only of replacement synthetic oil at $30 per barrel, but also of an elaborate conservation measure costing $12 per barrel saved and of solar heat at $20 per barrel equivalent supplied. (The choice is real, but these numbers are purely illustrative.) Present policy, true to asymmetric form, would be to brand the conservation and solar options as "uneconomic"—that is, more costly than short-term $10 oil—but nevertheless to subsidize the synthetics plant to the tune of $20 per barrel, even though it is the dearest and least certain option available for replacing the cheap oil! (In fact, Congress has not yet agreed to the synthetic fuels subsidies, and at this writing none of the three approaches is being aggressively implemented in the U.S. or elsewhere, except perhaps for conservation in Sweden—though conservation and solar energy

[33] B. Hannon, *Science 189*:95–102 (1975).
[34] *Supra* note 29.
[35] *Supra* notes 30, 31, and 33.

are starting to receive some minor subsidies in an effort to compete with the far larger subsidies given to conventional fuels and electricity.)

This asymmetry, as glaring as any we noted earlier, is typical of our approach to energy investments today. Technologies that are complex, glamorous, and backed by powerful constituencies are given lavish subsidies, subventions, bailouts, and exemptions from paying their own environmental and social costs, while technologies that are simple, mundane, and less endowed with wealthy lobbyists are subjected to a far more rigorous set of economic tests and requirements. If we wish to be fair, and to reach sound decisions, we should subject all our options to the same tests, whether they be economic, environmental, social, or whatever. If we do so, and if the analysis in this book is correct, then the "hard" technologies considered in section 2.2 will come at the bottom of the list[36] while the conservation measures and the "soft" and transitional technologies of sections 2.4–2.6 will come at the top.

The last refuge of people who have most to lose from such rationality—promoters of nuclear power—is to insist that though people are ready to accept the plutonium economy, they are not prepared, and never will be prepared, to accept properly priced energy, well insulated roofs, efficient cars, and solar collectors. Clearly this is a political judgment on which reasonable, or at least committed, observers can differ. Perhaps people could not possibly understand why cheap energy isn't as cheap as it looks, but only means paying a much larger bill outside the energy budget and after the next election. Perhaps people really do believe that the kinds of pollution they can't see don't exist. Perhaps people really want to become far more dependent on utilities and giant energy corporations and to tolerate any degree of autocracy for the sake of being able to flick a convenient switch. My own view, though, is that people are much smarter than that, are well ahead of their governments, and are impatient to have an energy policy that makes sense.

Moreover, the objection that any change in present policy means intolerable social change begs the question of what social changes the successful execution of present policy would entail. Critics who say a soft energy path is unacceptable because it must change lifestyles are implying that they themselves favor no change in lifestyles, even over fifty years. This implies a static, zero growth

[36] Comptroller-General of the U.S., "An Evaluation of Proposed Federal Assistance For Financing Commercialization of Emerging Energy Technologies," EMD-76-10 (Washington, D.C.: U.S. General Accounting Office, 24 August 1976). This report finds that energy conservation, solar heating, tertiary oil recovery, municipal waste combustion, and some geothermal projects are more cost-effective investments, more deserving of federal support now, than coal synthetics.

economy with no technical or social progress—presumably not what they have in mind. What they probably mean is that they desire no change in certain highly selective patterns and rates of change in lifestyle that they consider agreeable for themselves and appropriate for other people. That is a very different matter. Rational discussion requires that we be both explicit and symmetric. If soft paths are to be construed as more than technical fixes, then the lifestyle changes they entail should be compared with those entailed by other possible futures (in particular, any errors of fact or logic in the analysis of the hard path given in this book should be pointed out). Both desired and undesired changes should be considered. If, on the other hand, one asserts that continuing present behavior, perhaps with minor improvements, will only increase benefits and decrease costs, including social costs, then one must show in detail any respects in which a soft path depending purely on technical fixes cannot do the same thing.

Changing any policy, even one that is plainly unworkable, is never easy. It entails doubt, conflict, trial, error, and hard work. Taking the initial steps toward a soft energy path, and following up to be sure they work, will not be easy—only easier than not taking them. But if wisely handled they can have enormous political appeal. Instead of trading off one constituency against another—unemployment versus inflation, economic growth versus environmental quality, inconvenience versus vulnerability—a soft path offers advantages for every constituency. As Chapter Two suggests, a soft path simultaneously offers jobs for the unemployed, capital for businesspeople, environmental protection for conservationists, enhanced national security for the military, opportunities for small business to innovate and for big business to recycle itself, exciting technologies for the secular, a rebirth of spiritual values for the religious, traditional virtues for the old, radical reforms for the young, world order and equity for globalists, energy independence for isolationists, civil rights for liberals, states' rights for conservatives.

Thus, though present policy is consistent with the perceived short-term interests of a few powerful institutions, a soft path is consistent with far more strands of convergent social change at the grass roots. It goes with, not against, our political grain. And it is compatible with innovations in a great many other areas of public policy that we ought to be making anyhow for other reasons, and that plainly have the country behind them. If we free some log jams of outmoded perceptions that stifle our present approach to the energy problem, we can release a flood of change for the better.

Perhaps our salvation will yet be that the basic issues in energy strategy, far from being too complex and technical for ordinary people to understand, are on the contrary too simple and political for experts to understand. We must concentrate on these simple yet powerful ideas, not only if we are to gain a fuller understanding of the consequences of choice, but also if we are to appreciate the very wide range of choices that is available. And here too we must be symmetrical: if we do not like some aspects of the soft energy path, we must consider whether we prefer the hard path and all its consequences. Robert Frost, in the poem that forms our bridge to the next chapter, had an unmentioned third choice—bushwhacking through the shrubbery—but if we have that choice too, nobody has yet discovered it. The soft and the hard energy paths, and a myriad variations on their themes, appear to be the only choices there are, and we must decide which we prefer.

THE ROAD NOT TAKEN

Two roads diverged in a yellow wood,
And sorry I could not travel both
And be one traveller, long I stood
And looked down one as far as I could
To where it bent in the undergrowth;

Then took the other, as just as fair,
And having perhaps the better claim,
Because it was grassy and wanted wear;
Though as for that the passing there
Had worn them really about the same,

And both that morning equally lay
In leaves no step had trodden black.
Oh, I kept the first for another day!
Yet knowing how way leads on to way,
I doubted if I should ever come back.

I shall be telling this with a sigh
Somewhere ages and ages hence:
Two roads diverged in a wood, and I—
I took the one less travelled by,
And that has made all the difference.

—ROBERT FROST

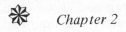 *Chapter 2*

Energy Strategy: The Road Not Taken?

2.1. OVERVIEW

Where are America's formal or de facto energy policies leading us? Where might we choose to go instead? How can we find out?

Addressing these questions can reveal deeper questions—and a few answers—that are easy to grasp, yet rich in insight and in international relevance. This chapter will seek to explore such basic concepts in energy strategy by outlining and contrasting two energy paths that the United States (or, by analogy, other countries) might follow over the next fifty years—long enough for the full implications of change to start to emerge. The first path resembles 1976-7 federal policy and is essentially an extrapolation of the recent past. It relies on rapid expansion of centralized high technologies to increase supplies of energy, especially in the form of electricity. The second path combines a prompt and serious commitment to efficient use of energy, rapid development of renewable energy sources matched in scale and in energy quality to end use needs, and special transitional fossil fuel technologies. This path, a whole greater than the sum of its parts, diverges radically from incremental past practices to pursue long-term goals. It does not try to wipe the slate clean, but rather to redirect our future efforts, taking advantage of the big energy systems we already have without multiplying them further.

Both paths, as will be argued, present difficult—but very different—problems. The first path is convincingly familiar, but the economic and sociopolitical problems lying ahead loom large, and eventually, perhaps, will prove insuperable. The second path, though it represents a shift in direction, offers many social, economic, and geopolitical advantages, including virtual elimination of nuclear proliferation from the world. It is important to recognize that the two paths are mutually exclusive. Because commitments to the first may foreclose the second, we must soon choose one or the other—before failure to stop nuclear proliferation has foreclosed both.[1]

2.2. HARD ENERGY PATHS

Most official proposals for future U.S. energy policy embody the twin goals of sustaining growth in energy consumption (assumed to be closely and causally linked to GNP and to social welfare) and of minimizing oil imports. The usual proposed solution is rapid expansion of three sectors: coal (mainly strip-mined, then made into electricity and synthetic fluid fuels); oil and gas (increasingly from Arctic and offshore wells); and nuclear fission (eventually in fast breeder reactors). All domestic resources, even naval oil reserves, are squeezed hard—in a policy that David Brower calls "Strength Through Exhaustion." Conservation, usually induced by price rather than by policy, is conceded to be necessary but it is given a priority more rhetorical than real. "Unconventional" energy supply is relegated to a minor role, its significant contribution postponed until past 2000. Emphasis is overwhelmingly on the short term. Long-term sustainability is vaguely assumed to be ensured by some eventual combination of fission breeders, fusion breeders, and solar electricity. Meanwhile, aggressive subsidies and regulations are used to hold down energy prices well below economic and prevailing international levels so that growth will not be seriously constrained.

Even over the first ten years (1976–1985), the supply enterprise typically proposed in such projections is impressive. Oil and gas extraction shift dramatically to offshore and Alaskan sources, with nearly 900 new oil wells offshore of the contiguous forty-eight

---

[1] In this chapter the proportions assigned to the components of the two paths are only indicative and illustrative. More exact computations, now being done by several groups in the United States and abroad, involve a level of technical detail which, though an essential next step, may deflect attention from fundamental concepts. This chapter will accordingly seek technical realism without rigorous precision or completeness. See Chapter Three for methodological discussion. Further technical details are given in later chapters and their citations. See also the independent but somewhat related analysis to be published in 1977 by the Union of Concerned Scientists (1208 Massachusetts Avenue, Cambridge, Massachusetts 02138) as the report of the UCS Energy Study.

states alone. Some 170 new coals mines open, extracting about 200 million tons per year each from eastern underground and strip mines, plus 120 million from western stripping. The nuclear fuel cycle requires over one hundred new uranium mines, a new enrichment plant, some forty fuel fabrication plants, three fuel reprocessing plants. The electrical supply system, more than doubling, draws on some 180 new 800 megawatt coal fired stations, over one hundred forty 1000 megawatt nuclear reactors, sixty conventional and over one hundred pumped storage hydroelectric plants, and over 350 gas turbines. Work begins on new industries to make synthetic fuels from coal and oil shale. At peak, just building (not operating) all these new facilities directly requires nearly 100,000 engineers, over 420,000 craftspeople, and over 140,000 laborers. Total indirect labor requirements are twice as great.[2]

This ten year spurt is only the beginning. The year 2000 finds us with 450 to 800 reactors (including perhaps eighty fast breeders, each loaded with 2.5 metric tons of plutonium), 500 to 800 huge coal-fired power stations, 1000 to 1600 new coal mines and some fifteen million electric automobiles. Massive electrification—"the most important attempt to modify the infrastructure of industrial society since the railroad"[3] is largely responsible for the release of waste heat sufficient in principle to warm the entire freshwater runoff of the contiguous forty-eight states by 34–49°F.[4] Mining coal and uranium, increasingly in the arid West, entails inverting thousands of communities and millions of acres, often with little hope of effective restoration. The commitment to a long-term coal economy many times the scale of today's makes the doubling of atmospheric carbon dioxide concentration early in the next century virtually unavoidable, with the prospect then or soon thereafter of substantial and perhaps irreversible changes in global climate.[5] Only the exact date of such changes is in question.

[2] The foregoing data are from M. Carasso et al., *The Energy Supply Planning Model*, PB-245 382 and PB-245 383 (Springfield, Virginia: National Technical Information Service, Bechtel Corp. report to the National Science Foundation [NSF], August 1975). The figures assume the production goals of the 1975 State of the Union Message. Indirect labor requirements are calculated by C.W. Bullard and D.A. Pilati, CAC Document 178 (September 1975), Center for Advanced Computation, University of Illinois at Urbana-Champaign. See Chapter 1.2.

[3] I.C. Bupp and R. Treitel, "The Economics of Nuclear Power: De Omnibus Dubitandum," 1976 (available from Professor Bupp, Harvard Business School).

[4] Computation concerning waste heat and projections to 2000 are based on data in the 1975 Energy Research and Development Administration Plan (ERDA-48).

[5] B. Bolin, "Energy and Climate," Secretariat for Future Studies (Fack, S-103

The main ingredients of such an energy future are roughly sketched in Figure 2-1. For the period up to 2000, this sketch is a composite of recent projections published by the Energy Research and Development Administration (ERDA), Federal Energy Administration (FEA), Department of the Interior, Exxon, and Edison Electric Institute. Minor and relatively constant sources, such as hydroelectricity, are omitted; the nuclear component represents nuclear heat, which is roughly three times the resulting nuclear electric output; fuel imports are aggregated with domestic production. Beyond 2000, the usual cutoff date of present projections, the picture has been extrapolated to the year 2025—exactly how is not important here—in order to show its long-term implications more clearly.[6]

2.3. WHY HARD PATHS FAIL

The flaws in this type of energy policy have been pointed out by critics in and out of government. For example, despite the intensive electrification—consuming more than half the total fuel input in 2000 and more thereafter—we are still short of gaseous and liquid fuels, acutely so from the 1980s on, because of slow and incomplete substitution of electricity for the two-thirds of fuel use that is now direct. Despite enhanced recovery of resources in the ground, shortages steadily deepen in natural gas—on which plastics and nitrogen fertilizers depend—and, later, in liquid fuel for the transport sector (half our oil now runs cars). Worse, at least half the energy growth never reaches the consumer because it is lost earlier in elaborate conversions in an increasingly inefficient fuel chain dominated by electricity generation (which wastes about two-thirds of the fuel) and

10 Stockholm); S.H. Schneider and R.D. Dennett, *Ambio 4*, 2:65-74 (1975); S.H. Schneider, *The Genesis Strategy* (New York: Plenum, 1976); W.W. Kellogg and S.H. Schneider, *Science 186*:1163-72 (1974); S.H. Schneider, *J. Atmos. Sci.* *32*:2060-66 (1975); W.W. Kellogg, "Effects of Human Activities on Global Climate," (Geneva: World Meterological Organization, October 1976), and "Global Influences of Mankind on the Climate," in J. Gribbin, ed., *Climate Change* (Cambridge, England: Cambridge University Press, 1977); R. Rotty, "Global Energy Demand and Related Climate Change," IEA(M)-75-3 (Oak Ridge: Institute for Energy Analysis, November 1975); W. Häfele, RR-76-1, IIASA (Laxenburg, Austria), pp. 15, 144-97. The CO_2 problem, as Häfele shows, is remarkably insensitive to technical (e.g., nuclear) assumptions if a high energy future is assumed.

[6] Figure 2-1 shows only *nonagricultural* energy. Yet the sunlight participating in photosynthesis in our harvested crops is comparable to our total use of nonagricultural energy, while the sunlight falling on *all* U.S. croplands and grazing lands is about twenty-five times the nonagricultural energy. By any measure, sunlight is the largest single energy input to the U.S. economy today.

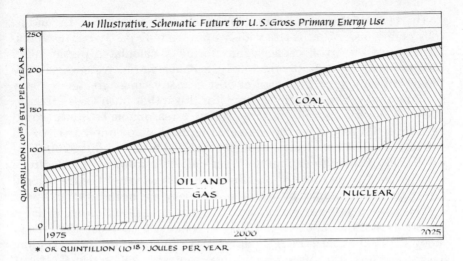

Figure 2-1.

coal conversion (which wastes about one-third). Thus in Britain since 1900, primary energy—the input to the fuel chain—has doubled while energy at the point of end use—the car, furnace, or machine whose function it fuels—has increased by only a half, or by a third per capita; the other half of the growth went to fuel the fuel industries, which are the largest energy consumers.

Among the most intractable barriers to implementing Figure 2-1 is its capital cost. In the 1960s, the total investment to increase a consumer's delivered energy supplies by the equivalent of one barrel of oil per day (about 67 kilowatts of heat) was a few thousand of today's dollars—of which, in an oil system, the wellhead investment in the Persian Gulf was and still is only a few hundred dollars. (The rest is transport, refining, marketing, and distribution.) The capital intensity of much new coal supply is still in this range. But such cheaply won resources can no longer stretch our domestic production of fluid fuels or electricity; and Figure 2-1 relies mainly on these, not on coal burned directly, so it must bear the full burden of increased capital intensity.

That burden is formidable. For the North Sea oilfields coming into production soon, the investment in the whole system is roughly $10,000 to deliver an extra barrel per day (constant 1976 dollars throughout); for U.S. frontier (Arctic and offshore) oil and gas in the 1980s it will be generally in the range from $10,000 to $25,000; for synthetic gaseous and liquid fuels made from coal, from $20,000 to $50,000 or more per daily barrel.

The scale of these capital costs is generally recognized in the industries concerned. What is less widely appreciated—partly because capital costs of electrical capacity are normally calculated per installed (not delivered) kilowatt and partly because whole system costs are rarely computed—is that capital cost is many times greater for new systems that make electricity than for those that burn fuels directly. For coal-electric capacity ordered today, a reasonable estimate would be about $170,000 for the delivered equivalent of one barrel of oil per day; for nuclear-electric capacity ordered today, about $200,000–$300,000. Thus, the capital cost per delivered kilowatt of electrical energy emerges as roughly one hundred times that of the traditional direct fuel technologies on which out society has been built.[7]

The capital intensity of coal conversion and, even more, of large electrical stations and distribution networks is so great that many analysts, such as the strategic planners of the Shell Group in London, have concluded that no major country outside the Persian Gulf can afford these centralized high technologies on a truly large scale, large enough to run a country. They are looking, in Monte Canfield's phrase, like future technologies whose time has passed.

Relying heavily on such technologies, the 1976–1985 energy program proposed in the January 1975 State of the Union Message turns out to cost over $1 trillion (in 1976 dollars) in initial investment, of which about 70 to 80 percent would be for new rather than replacement plants.[8] The latter figure corresponds to about three-fourths of cumulative net private domestic investment (NPDI) over the decade (assuming that NPDI remains 7 percent of gross national product and that GNP achieves real growth of 3.5 percent per year despite the adverse effects of the energy program on other investments). In contrast, the energy sector has recently required only one-fourth of NPDI. Diverting to the energy sector not only this hefty share of discretionary investment but also about two-thirds of all the rest would deprive other sectors that have their own cost escalation problems and their own vocal constituencies. A powerful political response could be expected. And this capital burden is not temporary; further up the curves of Figure 2–1 it tends to increase, and much of what might have been thought to be increased national wealth must be

---

[7] The capital costs for frontier fluids and for electrical systems can be readily calculated from the data base of the Bechtel model (see *supra* note 2). See Chapter Six for details.

[8] The Bechtel model, using 1974 dollars and assuming ordering in early 1974, estimates direct construction costs totaling $559 billion, including work that is in progress but not yet commissioned in 1985. Interest, design, and administration—but not land, nor escalation beyond the GNP inflation rate—bring the total to $743 billion. Including the cost of land, and correcting to a 1976 ordering date and 1976 dollars, is estimated by M. Carasso to yield over $1 trillion.

plowed back into the care and feeding of the energy system. Such long lead time, long payback time investments might also be highly inflationary.

Of the $1 trillion plus just cited, three-fourths would be for electrification. About 18 percent of the total investment could be saved just by reducing the assumed average 1976–1985 electrical growth rate from 6.5 to 5.5 percent per year.[9] Not surprisingly, the combination of disproportionate and rapidly increasing capital intensity, long lead times, and economic responses is already proving awkward to the electric utility industry, despite the protection of a 20 percent taxpayer subsidy on new power stations.[10] "Probably no industry," observes Bankers Trust Company, "has come closer to the edge of financial disaster." In many countries today an effective feedback loop is observable: large capital programs → poor cash flow → higher electricity prices → reduced demand growth → worse cash flow → increased bond flotation → increased debt-to-equity ratio, worse coverage, and less attractive bonds → poor bond sales → worse cash flow → higher electricity prices → reduced (even negative) demand growth and political pressure on utility regulators → overcapacity, credit pressure, and higher cost of money → worse cash flow, etc. This "spiral of impossibility," as Mason Willrich has called it, is exacerbated by most utilities' failure to base historic prices on the long-run cost of new supply: thus some must now tell their customers that the current dollar cost of a kilowatt-hour will treble by 1985, and that two-thirds of that increase will be capital charges for new plants. Moreover, experience abroad suggests that even a national treasury cannot long afford electrification: a New York State–like position is quickly reached, or too little money is left over to finance the energy *uses*, or both.

2.4. IMPROVING ENERGY EFFICIENCY

Summarizing a similar situation in Britain, Walter Patterson concludes: "Official statements identify an anticipated 'energy gap' which can be filled only with nuclear electricity; the data do not support any such conclusion, either as regards the 'gap' or as regards the capability of filling it with nuclear electricity." We have sketched one form of the latter argument; let us now consider the former.

[9] M. Carasso et al., *supra* note 2.
[10] E. Kahn et al., "Investment Planning in the Energy Sector," LBL-4479 (Berkeley, California: Lawrence Berkeley Laboratory, 1 March 1976). See also T.D. Mount & L.D. Chapman, in *Proceedings of the Workshop on Energy Demand* (22–23 May 1976), CP-76-1 (Laxenburg, Austria: IIASA, 1976), p. 164.

Despite the steeply rising capital intensity of new energy supply, forecasts of energy demand made as recently as 1972 by such bodies as the Federal Power Commission and the Department of the Interior wholly ignored both price elasticity of demand and energy conservation. The Chase Manhattan Bank in 1973 (and again in 1976) saw virtually no scope for conservation save by minor curtailments: the efficiency with which energy produced economic outputs was assumed to be optimal already. In 1977, some analysts still predict economic calamity if the United States does not continue to consume twice the combined energy total for Africa, the rest of North and South America, and Asia except Japan. But what have more careful studies taught us about the scope for doing better with the energy we have? Since we can't keep the bathtub filled because the hot water keeps running out, do we really (as Malcolm MacEwen asks) need a bigger water heater, or could we do better with a cheap, low technology plug?

There are two ways, divided by a somewhat fuzzy line, to do more with less energy. First, we can plug leaks and use thriftier technologies to produce exactly the same output of goods and services—and bads and nuisances—as before, substituting other resources (capital, design, management, care, etc.) for some of the energy we formerly used. When measures of this type use today's technologies, are advantageous today by conventional economic criteria, and have no significant effect on lifestyles, they are called "technical fixes."

In addition, or instead, we can make and use a smaller quantity or a different mix of the outputs themselves, thus to some degree changing (or reflecting ulterior changes in) our lifestyles. We might do this because of changes in personal values, rationing by price or otherwise, mandatory curtailments, or gentler inducements. Such "social changes" include carpooling, smaller cars, mass transit, bicycles, walking, opening windows, dressing to suit the weather, and extensively recycling materials. Technical fixes, on the other hand, include thermal insulation, heat pumps (devices like air conditioners that move heat around—often in either direction—rather than making it from scratch), more efficient furnaces and car engines, less overlighting and overventilation in commercial buildings, and recuperators for waste heat in industrial processes. Hundreds of technical and semitechnical analyses of both kinds of conservation have been done; in the last two years especially, much analytic progress has been made.

Theoretical analysis suggests that, in the long term, technical fixes *alone* in the United States could probably improve energy efficiency by a factor of at least three or four.[11] A recent review of specific practical measures cogently argues that with only those technical fixes that could be implemented by about the turn of the century, Americans could nearly double the efficiency with which they use energy.[12] If that is correct, economic activity could increase steadily with approximately constant primary energy use for the next few decades, thus stretching present energy supplies rather than having to add massively to them. One careful comparison shows that *after* correcting for differences of climate, geography, hydroelectric capacity, etc., Americans would still use about a third less energy than they do now if they were as efficient as the Swedes (who see much room for improvement in their own efficiency).[13] U.S. per capita energy intensity, too, is about twice that of West Germany in space heating, four times in transport.[14] Much of the difference is attributable to technical fixes.

Some technical fixes are already under way in the United States. Extensive new federal and state legislation is starting to be implemented. Many factories have cut tens of percent off their fuel cost per unit output, often with practically no capital investment. New 1976 cars averaged 27 percent better mileage than 1974 models. And there is overwhelming evidence that technical fixes are generally much cheaper than increasing energy supply, as well as quicker, safer, and of more lasting benefit. They are also better for secure, broadly based employment using existing skills. Most energy conservation measures and the shifts of consumption that they occasion are relatively labor intensive. Even making more energy-efficient home appliances is about twice as good for jobs as is building power stations: the latter is practically the least labor-intensive major investment in the whole economy.

The capital savings of conservation are particularly impressive. In the terms used above, the investments needed to *save* the equivalent of an extra barrel of oil per day are often zero to $3500, generally

---

[11] American Institute of Physics Conference Proceedings No. 25, *Efficient Use of Energy* (New York: AIP, 1975), summarized in *Physics Today*, August 1975.
[12] M. Ross and R.H. Williams, *Bull. Atom. Scient.* 32:9, 30–38 (November 1976) and *Technology Review*, in press (1977); see also L. Schipper, *Ann. Rev. Energy* 1:455–518 (1976); and R.H. Socolow, "The Coming Age of Conservation," *Ann. Rev. Energy* 2: in press (1977).
[13] L. Schipper and A.J. Lichtenberg, *Science 194*:1001–13 (1976).
[14] R.L. Goen and R.K. White, "Comparison of Energy Consumption Between West Germany and the United States" (Menlo Park, California: Stanford Research Institute, June 1975).

under $8000, and at most about $25,000—far less than the amounts needed to increase most kinds of energy supply. Indeed, to use energy efficiently in new buildings, especially commercial ones, the additional capital cost is often *negative*: savings on heating and cooling equipment more than pay for the other modifications.

To take one major area of potential saving, technical fixes in new buildings—almost anywhere in the world—can save 50 percent or more in office buildings and 80 percent or more in some new U.S. houses.[15] A recent American Institute of Architects study concludes that, by 1990, improved design of new buildings and modification of old ones could save a third of our current *total* U.S. energy use—and save money too. The payback time would be only half that of the alternative investment in increased energy supply, so the same capital could be used twice over.

A second major area lies in "cogeneration," or the generating of electricity as a by-product of the process steam normally produced in many industries. A Dow study chaired by Paul McCracken reports that by 1985 U.S. industry could meet approximately half its own electricity needs (compared to about a seventh today) by this means. Such cogeneration would save $20–50 billion in investment, save fuel equivalent to two to three million barrels of oil per day, obviate the need for more than fifty large reactors, and (with flattened utility rates) yield at least 20 percent pretax return on marginal investment while reducing the price of electricity to consumers.[16] Another mea-

---

[15] A.D. Little, Inc., "An Impact Assessment of ASHRAE Standard 90–75," report to FEA, C-78309, December 1975; J.E. Snell et al. (National Bureau of Standards), "Energy Conservation in Office Buildings: Some United States Examples," International CIB Symposium on Energy Conservation in the Built Environment (Building Research Establishment, Garston, Watford, England), April 1976 (Hornby, Lancs.: Construction Press Ltd., 1976); Owens-Corning-Fiberglas (Toledo, Ohio), "The Arkansas Story," 1975; E. Hirst, *Science 194*: 1247 (1976); recent publications of the American Institute of Architects (Washington, D.C.); Dubin-Mindell-Bloome Associates, *A Study of Existing Energy Usage on Long Island and the Impact of Energy Conservation, Solar Energy, Total Energy and Wind Systems on Future Requirements* (New York, 31 October 1975).

[16] P.W. McCracken et al., *Industrial Energy Center Study*, Dow Chemical Co. et al. report to NSF, PB-243 824, NTIS, June 1975. Two important studies published more recently have examined a wider range of sizes and types of cogeneration systems and have concluded that the Dow report substantially underestimates the potential: S.E. Nydick et al., "A Study of Inplant Electric Power Generation in the Chemical, Petroleum Refining, and Paper and Pulp Industries," Thermo Electron Corporation report to FEA, PB-255-658 and -659, NTIS, May 1976; and R.H. Williams, "The Potential for Electricity Generation as a Byproduct of Industrial Steam Production in New Jersey," report to N.J. Cabinet Energy Committee (Princeton, New Jersey: Center for Environmental Studies, Princeton University, 21 June 1976).

sure of the potential is that cogeneration, whose contribution to U.S. electricity supply has fallen from about 15 percent in 1950 to about 4 percent today, still supplies about 12 percent in West Germany. Cogeneration and more efficient use of electricity could together reduce U.S. use of electricity by a third and central station generation by 60 percent.[17] Like district heating (distribution of waste heat as hot water via insulated pipes to heat buildings), U.S. cogeneration is held back only by institutional barriers. Yet these are smaller than those that were overcome when the present utility industry was established.

So great is the scope for technical fixes now that the U.S. could spend several hundred billion dollars on them initially plus several hundred million dollars per day—and still save money compared with increasing the supply! And one would still have the fuel (without the environmental and geopolitical problems of getting and using it). The barriers to far more efficient use of energy are not technical, nor in any fundamental sense economic. So why do we stand here, confronted, as Pogo said, by insurmountable opportunities?

The answer—apart from poor information and ideological antipathy and rigidity—is a wide array of institutional barriers, including more than 3000 conflicting and often obsolete building codes, an innovation-resistant building industry, lack of mechanisms to ease the transition from kinds of work that we no longer need to kinds we do need, opposition by strong unions to schemes that would transfer jobs from their members to larger numbers of less "skilled" workers, promotional utility rate structures, fee structures giving building engineers a fixed percentage of prices of heating and cooling equipment they install, inappropriate tax and mortgage policies, conflicting signals to consumers, misallocation of conservation's costs and benefits (builders versus buyers, landlords versus tenants, etc.), imperfect access to capital markets, fragmentation of government responsibility, etc.

Though economic answers are not always right answers, properly using the markets we have (see Chapter 1.5) may be the greatest single step we could take toward a sustainable, humane energy future. The sound economic principles we need to apply include flat (even inverted) utility rate structures rather than discounts for large users, pricing energy according to what extra supplies will cost in the long run ("long-run marginal cost pricing"), removing subsidies, assessing the total costs of energy-using purchases over their whole

[17] Ross and Williams, *supra* note 12. A further 5 quad saving in U.S. primary energy through currently economic combined heat and power stations and district heating grids—which could reach at least half the U.S. population—is calculated by J. Karkheck et al., *Science 195*:948–55 (1977).

operating lifetimes ("life cycle costing"), counting the costs of complete energy systems including all support and distribution systems, properly assessing and charging environmental costs, valuing assets by what it would cost to replace them, discounting appropriately, and encouraging competition through antitrust enforcement (including at least horizontal divestiture of giant energy corporations).

Such practicing of the market principles we preach could go very far to help us use energy efficiently and get it from sustainable sources. But just as clearly, there are things the market cannot do, like reforming building codes or utility practices. And whatever our means, there is room for differences of opinion about how far we can achieve the great theoretical potential for technical fixes. How far might we instead choose, or be driven to, some of the "social changes" mentioned earlier?

There is no definitive answer to this question—though it is arguable that if we are not clever enough to overcome the institutional barriers to implementing technical fixes, we shall certainly not be clever enough to overcome the more familiar but more formidable barriers to increasing energy supplies. My own view of the evidence is, first, that North Americans are adaptable enough to use technical fixes *alone* to double, in the next few decades, the amount of social benefit wrung from each unit of end-use energy; and second, that value changes that could either replace or supplement those technical changes are also occurring rapidly. If either of these views is right, or if both are partly right, North Americans should be able to double end-use efficiency by the turn of the century or shortly thereafter, with minor or no changes in lifestyles or values save increasing comfort for modestly increasing numbers, then over the period 2010-2040 to shrink per capita primary energy use to perhaps a third or a quarter of today's.[18] (The former would reach the per capita level of the wasteful, but hardly troglodytic, French; the latter, the level of the New Zealanders or the 1970 Swiss. Even in the case of fourfold shrinkage, the resulting society could be instantly recognizable to a visitor from the 1960s and need in no sense be a pastoralist's utopia—

[18] A calculation for Canada supports this view: A.B. Lovins, *Conserver Society Notes* (Ottawa: Science Council of Canada, May/June 1976), pp. 3-16. Technical fixes already approved in principle by the Canadian Cabinet should hold approximately constant until 1990 the energy required for the transport, commercial, and house-heating sectors; sustaining similar measures to 2025 is estimated to shrink per capita primary energy to about half today's level. Plausible social changes are estimated to yield a further halving. The Canadian and U.S. energy systems have rather similar structures. The potential for increasing end-use efficiency is considerably less in Europe than in North America: a doubling might be expected in Europe over the next fifty years or so, rather than the North American quadrupling.

though that option would remain open to those who may desire it.

The long-term mix of technical fixes with structural and value changes in work, leisure, agriculture, and industry will require much trial and error. It will take many years to make up our diverse minds about. It will not be easy—merely easier than not doing it. Meanwhile, it is easy only to see what not to do.

If one assumes that by resolute technical fixes and modest social innovation North Americans can double their end-use efficiency by shortly after 2000, then they could be twice as affluent as now with today's level of energy use, or as affluent as now while using only half the end-use energy they use today. Or they might be somewhere in between—significantly more affluent (and equitable) than today but with less end-use energy.

Many analysts now regard modest, zero, or negative growth in the rate of energy use in industrial countries as a realistic long-term goal. Present annual U.S. primary energy demand, for example, is about seventy-five quadrillion BTU ("quads"), and most official projections for 2000 envisage growth to 130-170 quads. However, recent work at the Institute for Energy Analysis, Oak Ridge, under the direction of Dr. Alvin Weinberg, suggests that standard projections of energy demand are far too high because they do not take into account changes in demographic and economic trends. In June 1976 the institute considered that with a conservation program far more modest than that contemplated in this article, the likely range of U.S. primary energy demand in the year 2000 would be about 101-126 quads, with the lower end of the range more probable and end-use energy being about 60-65 quads, much less than is considered here. In early 1977, Drs. R.H. Williams and F. von Hippel of Princeton University likewise showed in their testimony to the Nuclear Regulatory Commission's GESMO hearings that 112 quads in 2000 could be considered a "business-as-usual" projection assuming only the conservation measures already enacted, with further modest conservation yielding only eighty-nine quads. And, at the further end of the spectrum, projections of U.S. primary energy for 2010 being seriously studied by the Committee on Nuclear and Alternative Energy Systems, a major U.S. National Research Council study due to report in mid-1977, ranged as low as about seventy quads (fifty-four quads of fuels plus sixteen of solar energy), with an even lower figure (forty to fifty quads total primary energy) being examined.

As the basis for a coherent alternative to the path shown in Figure 2-1, Figure 2-2 sketches a primary energy demand of about ninety-five quads for 2000—a value that the above data suggest is by no means the lowest that could be realistically considered. Total energy

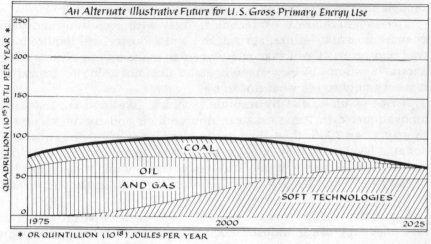

Figure 2-2.

demand would gradually decline thereafter as inefficient buildings, machines, cars, and energy systems are slowly modified or replaced. Let us now explore the other ingredients of such a path—starting with the "soft" supply technologies which, spurned in Figure 2-1 as insignificant, now assume great importance.

2.5. SOFT ENERGY TECHNOLOGIES

There exists today a body of energy technologies that have certain specific features in common and that offer great technical, economic, and political attractions, yet for which there is no generic term. For lack of a more satisfactory term, I shall call them "soft" technologies: a textural description, intended to mean not vague, mushy, speculative, or ephemeral, but rather flexible, resilient, sustainable, and benign. Energy paths dependent on soft technologies, illustrated in Figure 2-2, will be called "soft" energy paths, as the "hard" technologies sketched in Chapter 2.2 constitute a "hard" path (in both senses). The distinction between hard and soft energy paths rests not on how much energy is used, but on the technical and sociopolitical *structure* of the energy system, thus focusing our attention on consequent and crucial political differences.

In Figure 2-2, then, the social structure is significantly shaped by the rapid deployment of soft technologies. These are defined by five characteristics:

1. They rely on renewable energy flows that are always there whether

we use them or not, such as sun and wind and vegetation: on energy income, not on depletable energy capital.

2. They are diverse, so that as a national treasury runs on many small tax contributions, so national energy supply is an aggregate of very many individually modest contributions, each designed for maximum effectiveness in particular circumstances.

3. They are flexible and relatively low technology—which does not mean unsophisticated, but rather, easy to understand and use without esoteric skills, accessible rather than arcane (see Chapter Nine).

4. They are matched in *scale* and in geographic distribution to end use needs, taking advantage of the free distribution of most natural energy flows.

5. They are matched in *energy quality* to end-use needs: a key feature that deserves immediate explanation.

People do not want electricity or oil, nor such economic abstractions as "residential services," but rather comfortable rooms, light, vehicular motion, food, tables, and other real things. Such end-use needs can be classified by the physical nature of the task to be done (see Chapter Four). In the United States today, about 58 percent of all energy at the point of end use is required as heat, split roughly 23–35 between temperatures above and below the boiling point of water. (In Western Europe the low temperature heat alone is often a half of all end-use energy.) Another 38 percent of all U.S. end use energy provides mechanical motion: 31 percent in vehicles, 3 percent in pipelines, 4 percent in industrial electric motors. The rest, a mere 4 percent of delivered energy, represents *all* lighting, electronics, telecommunications, electrometallurgy, electrochemistry, arc welding, electric motors in home appliances and in railways, and similar end uses that now *require* electricity.

Some 8 percent of all U.S. energy end use, then, and similarly little abroad (see Chapter 4), requires electricity for purposes other than low temperature heating and cooling. Yet, since we actually use electricity for many such low grade purposes, it now meets 13 percent of U.S. end-use needs—and its generation consumes 29 percent of U.S. fossil fuels. A hard energy path would increase this 13 percent figure to 20–40 percent (depending on assumptions) by the year 2000, and far more thereafter. But this is wasteful because the laws of physics require, broadly speaking, that a power station change three units of fuel into two units of almost useless waste heat plus one unit of electricity. This electricity can do more difficult kinds of work than can the original fuel, but unless this extra quality and versatility are used to advantage, the costly process of upgrading the fuel—and losing two-thirds of it—is all for naught.

Plainly we are using premium fuels and electricity for many tasks for which their high energy quality is superfluous, wasteful, and expensive, and a hard path would make this inelegant practice even more common. Where we want only to create temperature differences of tens of degrees, we should meet the need with sources whose potential is tens or hundreds of degrees, not with a flame temperature of thousands or a nuclear reaction temperature equivalent to trillions—like cutting butter with a chainsaw.

For some applications, electricity is appropriate and indispensable: electronics, smelting, subways, most lighting, some kinds of mechanical work, and a few more. But these uses are already oversupplied, and for the other, dominant, uses remaining in our energy economy this special form of energy cannot give us our money's worth (in many parts of the United States today it already costs $50–120 per barrel equivalent). Indeed, in probably no industrial country today can additional supplies of electricity be used to thermodynamic advantage that would justify their high cost in money and fuels.

So limited are the U.S. end uses that really require electricity that by applying careful technical fixes to them we could reduce their 8 percent total to about 5 percent (mainly by reducing commercial overlighting), whereupon we could probably cover all those needs with present U.S. hydroelectric capacity plus the cogeneration capacity available in the mid to late 1980s.[19] Thus an affluent industrial economy could advantageously operate with no central power stations at all! In practice we would not necessarily want to go that far, at least not for a long time; but the possibility illustrates how far we are from supplying energy only in the quality needed for the task at hand.

Just as soft technologies' matching of energy quality to end–use needs virtually eliminates the costs and losses of secondary energy conversion, so the appropriate scale (see Chapter Five) of soft technologies can virtually eliminate the costs and losses of energy distribution. Matching scale to end uses can indeed achieve at least five important types of economies (see Chapter Five) not available to larger, more centralized systems. The first type is reduced and shared overheads. At least half your electricity bill is fixed distribution costs to pay the overheads of a sprawling energy system: transmission lines, transformers, cables, meters and people to read them, planners,

---

[19] The scale of potential conservation in this area is given in Ross and Williams, *supra* note 12; the scale of potential cogeneration capacity is from McCracken et al., and from Nydick et al., *supra* note 16.

headquarters, billing computers, interoffice memos, advertising agencies. For electrical and some fossil fuel systems, distribution accounts for more than half of total capital cost, and administration for a significant fraction of total operating cost. Local or domestic energy systems can reduce or even eliminate these infrastructure costs. The resulting savings can far outweigh the extra costs of the dispersed maintenance infrastructure that the small systems require, particularly where that infrastructure already exists or can be shared (e.g., plumbers fixing solar heaters as well as sinks).

Small scale brings further savings by virtually eliminating distribution losses, which are cumulative and pervasive in centralized energy systems (particularly those using high quality energy). Small systems also avoid direct diseconomies of scale, such as the frequent unreliability of large units and the related need to provide instant "spinning reserve" capacity on electrical grids to replace large stations that suddenly fail. Small systems with short lead times greatly reduce exposure to interest, escalation, and mistimed demand forecasts—major indirect diseconomies of large scale.

The fifth type of economy available to small systems arises from mass production. Consider, as Henrik Harboe suggests, the 100-odd million cars in the U.S. In round numbers, each car probably has an average cost of less than $4000 and a shaft power over 100 kilowatts (134 horsepower). Presumably a good engineer could build a generator and upgrade an automobile engine to a reliable, 35 percent efficient diesel at no greater total cost, yielding a mass-produced diesel generator unit costing less than $40 per kW. In contrast, the motive capacity in U.S. central power stations—currently totaling about one-fortieth as much as in U.S. cars—costs perhaps ten times more per kW, partly because it is not mass produced. This is not to argue for the widespread use of diesel generators; rather, to suggest that if we could build power stations the way we build cars, they would cost at least ten times less than they do, but we can't because they're too big. In view of this scope for mass-producing small systems, it is not surprising that at least one European car maker hopes to go into the wind machine and heat pump business. Such a market can be entered incrementally, without the billions of dollars' investment required for, say, liquefying natural gas or gasifying coal. It may require a production philosophy oriented toward technical simplicity, low replacement cost, slow obsolescence, high reliability, high volume, and low markup; but these are familiar concepts in mass production. Industrial resistance would presumably melt when—as with pollution abatement equipment—the scope for profit was perceived.

This is not to say that all energy systems need be at domestic scale. The object is to crack nuts with nutcrackers and drive pilings with triphammers, not the reverse: to use the most appropriately scaled tool for the job and so minimize costs, including social costs. In some cases this will require big systems, chiefly the existing hydroelectric dams. In most cases the scale needed will be smaller. For example, the medium scale of urban neighborhoods and rural villages offers fine prospects for solar collectors—especially for adding collectors to existing buildings of which some (perhaps with large flat roofs) can take excess collector area while others cannot take any. They could be joined via communal heat storage systems, saving on labor cost and on heat losses. The costly craftwork of remodeling existing systems—"backfitting" or "retrofitting" idiosyncratic houses with individual collectors—could thereby be greatly reduced. Despite these advantages, medium-scale solar technologies are currently receiving little attention apart from a condominium village project in Vermont sponsored by the Department of Housing and Urban Development and the one hundred dwelling unit Méjannes-le-Clap project in France.

The schemes that dominate ERDA's solar research budget—such as making electricity from huge collectors in the desert, or from temperature differences in the oceans, or from Brooklyn Bridge–like satellites in outer space—do not satisfy our criteria, for they are ingenious high technology ways to supply energy in a form and at a scale inappropriate to most end–use needs. Not all solar technologies are soft. Nor, for the same reason, is nuclear fusion a soft technology.[20] But many genuine soft technologies are now available and are now economic. What are some of them?

Solar heating and, imminently, cooling head the list. They are incrementally cheaper than electric heating, and far more inflation-proof, practically anywhere in the world.[21] In the United States

[20] Assuming (which is still not certain) that controlled nuclear fusion works, it will almost certainly be more difficult, complex, and costly—though safer and perhaps more permanently fueled—than fast breeder reactors. See W.D. Metz, *Science 192*:1320-23 (1976); *193*:38-40, 76 (1976); and *193*:307-309 (1976). But for three reasons we ought not to pursue fusion. First, it generally produces copious fast neutrons that can and probably would be used to make bomb materials. Second, if it turns out to be rather "dirty," as most fusion experts expect, we shall probably use it anyway, whereas if it is clean, we shall so overuse it that the resulting heat release will alter global climate: we should prefer energy sources that give us enough for our needs while denying us the excesses of concentrated energy with which we might do mischief to the earth or to each other. Third, fusion is a clever way to do something we don't really want to do, namely to find *yet another* complex, costly, large-scale, centralized, high technology way to make electricity—all of which goes in the wrong direction.

[21] Partial or total solar heating is attractive and is being demonstrated even in

(with fairly high average sunlight levels), they are cheaper than present electric heating virtually anywhere, cheaper than oil heat in many parts, and cheaper than gas and coal in some. Even in the least favorable parts of the continental United States, far more sunlight falls on a typical building than is required to heat and cool it without supplement; whether this is considered economic depends on how the accounts are done.[22] The difference in solar input between the most and least favorable parts of the lower forty-nine states is generally less than twofold, and in cold regions, the long heating season can improve solar economics.

Ingenious ways of backfitting existing urban and rural buildings (even large commercial ones) or their neighborhoods with efficient and exceedingly reliable solar collectors (Chapter 7.4) are being rapidly developed in both the private and public sectors. In some recent projects, the lead time from ordering to operation has been only a few months. Good solar hardware, often modular, is going into pilot or full scale production over the next few years, and will increasingly be integrated into buildings as a multipurpose structural element, thereby sharing costs. Such firms as Philips, Honeywell, Revere, Pittsburgh Plate Glass, and Owens-Illinois, plus many dozens of smaller firms, are applying their talents, with rapid and accelerating effect, to reducing unit costs and improving performance. Some novel types of very simple collectors with far lower costs also show promise in current experiments. Indeed, solar hardware per se is necessary only for backfitting existing buildings. If we build new buildings properly in the first place, they can use "passive" solar collectors—large south windows or glass-covered black south walls—rather than special collectors. If we did this to all new houses in the next twelve years, we would save about as much energy as we expect to recover from the Alaskan North Slope.[23]

cloudy countries approaching the latitude of Anchorage, such as Denmark and the Netherlands (International CIB Symposium, *supra* note 15) and Britain (*Solar Energy: A U.K. Assessment*, International Solar Energy Society, London, May 1976). See also Chapter Seven, note 38.

[22] Solar heating cost is traditionally computed microeconomically for a consumer whose alternative fuels are not priced at long-run marginal cost (see, e.g., G. Bennington et al.'s MITRE study M76/79, "An Economic Analysis of Solar Water and Space Heating" [November 1976], announced by the ERDA Solar Division on 29 December 1976 [release 76-376], which also assumes unrealistically high solar costs and 100% backup capacity). Another method would be to compare the total cost (capital and life cycle) of the solar system with the total cost of the other complete systems that would otherwise have to be used in the long run to heat the same space. On that basis, 100 percent solar heating, even with twice the capital cost of two-thirds or three-fourths solar heating, is almost always advantageous. See Chapter Eight, and H.A. Bethe and A.B. Lovins, exchange of letters, *Foreign Affairs*, April 1977.

[23] R.W. Bliss, *Bull. Atom. Scient. 32*:3, 32–40 (March 1976).

Second, exciting developments in the conversion of agricultural, forestry, and urban wastes to methanol and other liquid and gaseous fuels (Chapter 7.3) now offer practical, economically interesting technologies sufficient to run an efficient U.S. transport sector.[24] Some bacterial and enzymatic routes under study look even more promising, but presently proved processes already offer sizable contributions without the inevitable climatic constraints of fossil fuel combustion. Organic conversion technologies must be sensitively integrated with agriculture and forestry so as not to deplete the soil; most current methods seem suitable in this respect, though they may change the farmer's priorities by making his whole yield of biomass (vegetable matter) salable.

The required scale of organic conversion can be estimated. Each year the U.S. beer and wine industry, for example, microbiologically produces 5 percent as many gallons (not all alcohol, of course) as the U.S. oil industry produces gasoline. Gasoline has 1.5 to 2 times the fuel value of alcohol per gallon. Thus a conversion industry roughly ten to fourteen times the physical scale (in gallons of fluid output per year) of U.S. cellars and breweries, albeit using different processes, would produce roughly one-third of the present gasoline requirements of the United States. If one assumes a transport sector with three times today's average efficiency—a reasonable estimate for early in the next century—then the whole of the transport needs could be met by organic conversion. The scale of effort required does not seem unreasonable, since it would replace in function half the present refinery capacity.

Additional soft technologies include wind hydraulic systems (especially those with a vertical axis), which already seem likely in many design studies to compete with nuclear power in much of North America and Western Europe. But wind (see Chapter 7.2) is not restricted to making electricity: it can heat, pump, heat-pump, or compress air. Solar process heat, too, is coming along rapidly (Chapter 7.4) as we learn to use the 5800°C potential of sunlight (much hotter than a boiler). Finally, high and low temperature solar collectors, organic converters, and wind machines can form symbiotic hybrid combinations more attractive than the separate components.

Energy storage is often said to be a major problem of energy income technologies. But this "problem" is largely an artifact of trying to recentralize, upgrade and redistribute inherently diffuse energy flows. Directly storing sunlight or wind—or, for that matter, electricity from any source—is indeed difficult on a large scale. But it is

[24] A.D. Poole and R.H. Williams, *Bull. Atom. Scient. 32*:5, 48–58 (May 1976).

easy if done on a scale and in an energy quality matched to most end use needs. Daily, even seasonal, storage of low and medium temperature heat at the point of use is straightforward with water tanks, rock beds, or perhaps fusible salts. Neighborhood heat storage is even cheaper. In industry, wind-generated compressed air can easily (and, with due care, safely) be stored to operate machinery: the technology is simple, cheap, reliable, and highly developed. (Some European cities even used to supply compressed air as a standard utility.) Installing pipes to distribute hot water (or compressed air) tends to be considerably cheaper than installing equivalent electric distribution capacity. Hydroelectricity is stored behind dams, and organic conversion yields readily stored liquid and gaseous fuels. On the whole, therefore, energy storage is much less of a problem in a soft energy economy than in a hard one.

Recent research suggests that a largely or wholly solar economy can be constructed in the United States with straightforward soft technologies that are now demonstrated and now economic or nearly economic.[25] Such a conceptual exercise does not require "exotic" methods such as sea-thermal, hot-dry-rock geothermal, cheap (perhaps organic) photovoltaic, or solar-thermal electric systems. If developed, as some probably will be, these technologies could be convenient, but they are in no way essential for an industrial society operating solely on energy income.

Figure 2-2 shows a plausible and realistic growth pattern, based on several detailed assessments (see Chapter Seven and section 5.4 of Chapter Five), for soft technologies given aggressive support. The useful output from these technologies would overtake, starting in the 1990s, the output of nuclear electricity shown in even the most sanguine federal estimates. For illustration, Figure 2-2 shows soft technologies meeting virtually all energy needs in 2025, reflecting a judgment that a completely soft supply mix is practicable in the long run, with or without the 2000-2025 energy shrinkage shown. Though most technologists who have thought seriously about the matter will concede it conceptually, some may be uneasy about the details. Obviously the sketched curve is not definitive, for although the general direction of the soft path must be shaped soon, the details of the energy economy in 2025 would not be committed in this century. To a large extent, therefore, it is enough to ask yourself whether Figure 2-1 or 2-2 seems preferable in the 1975-2000 period.

[25] For examples, see the Canadian computations in A.B. Lovins, *Conserver Society Notes* (*supra* note 18), Bent Sørensen's Danish estimates in *Science 189*:255-60 (1975), and, as a useful data base, the forthcoming estimates by the Union of Concerned Scientists (*supra* note 1).

A simple comparison, shown schematically in Figure 2-3, may help. Roughly half, perhaps more, of the gross primary energy being produced in the hard path in 2025 is lost in conversions. A further appreciable fraction is lost in distribution. Delivered end–use energy is thus not vastly greater than in the soft path, where conversion and distribution losses have been all but eliminated. (What is lost can often be used locally for heating, and is renewable, not depletable.) But the soft path makes each unit of end–use energy perform several times as much social function as it would have done in the hard path; so in a conventional sense, social welfare in the soft path in 2025 is substantially greater than in the hard path at the same date.

2.6. TRANSITIONAL ENERGY TECHNOLOGIES

To fuse into a coherent strategy the benefits of energy efficiency and of soft technologies, we need one further ingredient: transitional technologies that use fossil fuels briefly and sparingly to build a bridge to the energy income economy of 2025, conserving those fuels—especially oil and gas—for petrochemicals (ammonia, plastics, etc.) and leaving as much as possible in the ground for emergency use only.

Some transitional technologies have already been mentioned under the heading of conservation—specifically, cogenerating electricity from existing industrial steam and using existing waste heat for district heating. Given such measures, increased end–use efficiency, and the rapid development of biomass alcohol as a portable liquid fuel, the principal short-and medium-term problem becomes, not a shortage of electricity or of portable liquid fuels, but a shortage of clean sources of heat. It is above all the sophisticated use of coal, chiefly at modest scale, that needs development. Technical measures to permit the highly efficient use of this widely available fuel would be the most valuable transitional technologies.

Neglected for so many years, coal technology (see Chapter 7.1) is now experiencing a virtual revolution. We are developing supercritical gas extraction, flash hydrogenation, flash pyrolysis, panel-bed filters, and similar ways to use coal cleanly at essentially any scale and to cream off valuable liquids and gases as premium fuels before burning the rest. These methods largely avoid the costs, complexity, inflexibility, technical risks, long lead times, large scale, and tar formation of the traditional processes that now dominate our research.

Perhaps the most exciting current development is the so-called fluidized bed system for burning coal (or virtually any other combustible material). Fluidized beds are simple, versatile devices that

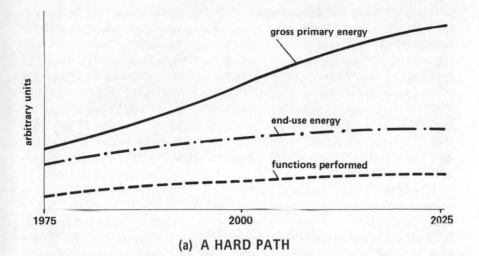

(a) A HARD PATH

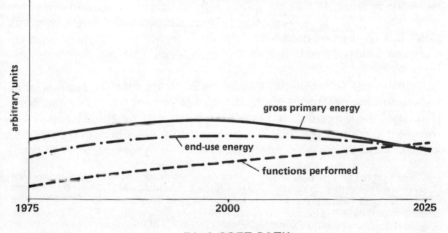

(b) A SOFT PATH

Figure 2-3. Schematic Sketch of Gross Primary Energy, End-Use Energy, and Quantity of Functions Performed by End-Use Energy in Hard and Soft Energy Paths.

add the fuel a little at a time to a much larger mass of small, inert, red-hot particles—sand or ceramic pellets—kept suspended as an agitated fluid by a stream of air continuously blown up through it from below. The efficiency of combustion, of other chemical reactions (such as sulfur removal), and of heat transfer is remarkably high because of the turbulent mixing and large surface area of the particles. Fluidized beds have long been used as chemical reactors and for burning trash, but are now ready to be commercially applied to raising steam and operating turbines. In one system currently available from Stal-Laval Turbin AB of Sweden, eight off the shelf, 70 megawatt gas turbines powered by fluidized bed combusters, together with district-heating networks and heat pumps, would heat as many houses as a $1 billion plus coal gasification plant, but would use only two-fifths as much coal, cost a half to two-thirds as much to build, and burn more cleanly than a normal power station with the best modern scrubbers.[26]

Fluidized bed boilers and turbines can power giant industrial complexes, especially for cogeneration, and are relatively easy to backfit into old municipal power stations. Scaled down, a fluidized bed can be a tiny household device—clean, strikingly simple, and flexible—that can replace an ordinary furnace or grate and can recover combustion heat with an efficiency over 80 percent.[27] At medium scale, such technologies offer versatile boiler backfits and improve heat recovery in flues. With only minor modifications they can burn practically any fuel. It is essential to commercialize all these systems now—not to waste a decade on highly instrumented but noncommercial pilot plants constrained to a narrow, even obsolete design philosophy.[28]

Transitional technologies can be built at appropriate scale so that soft technologies can be plugged into the system later. For example, if district heating uses hot water tanks on a neighborhood scale, those tanks can in the long run be heated by neighborhood solar collectors, wind-driven heat pumps, a factory, a pyrolyzer, a geothermal

[26] The system and its conceptual framework are described in several papers by H. Harboe, Managing Director, Stal-Laval (G.B.) Ltd., 41-7 Strand, London WC.2: "District Heating and Power Generation," 14 November 1975; "Advances in Coal Combustion and Its Applications," 20 February 1976; "Pressurized Fluidized Bed Combustion with Special Reference to Open Gas Turbines" (with C.W. Maude), May 1976. See also K.D. Kiang et al., "Fluidized-Bed Combustion of Coals," GFERC/IC-75/2 (CONF-750586), ERDA, May 1975.

[27] Small devices were pioneered by the late Professor Douglas Elliott. His associated firm, Fluidfire Development, Ltd. (Netherton, Dudley, W. Midlands, England), has sold many dozens of units for industrial heat treatment or heat recuperation. Field tests of domestic packaged fluidized bed boilers are in progress in the Netherlands and planned in Montana.

[28] In late 1977, Enköping, Sweden, expects to commission a 25 megawatt fluidized bed boiler for its district heating system. New reviews at the Institute for Energy Analysis, the U.S. House of Representatives Committee on Science and Technology, and elsewhere confirm fluidized beds' promise of rapid benefits without massive research programs. The Tennessee Valley Authority has announced plans to build a 200 MW fluidized bed boiler with or without ERDA's help.

well, or whatever else becomes locally available—offering flexibility that is not possible at today's excessive scale.

Both transitional and soft technologies are worthwhile industrial investments that can recycle moribund capacity and underused skills, stimulate exports, and give engaging problems to innovative technologists. Though neither glamorous nor militarily useful, these technologies are socially effective—especially in poor countries that need such scale, versatility and simplicity even more than rich countries do.

Properly used, coal, conservation, and soft technologies together can squeeze the "oil and gas" wedge in Figure 2-2 from both sides—so far that most of the frontier extraction and medium-term imports of oil and gas become unnecessary and conventional resources are greatly stretched. Coal can fill the real gaps in the fuel economy with only a temporary and modest (less than twofold at peak) expansion of mining, not requiring the enormous infrastructure and social impacts implied by the scale of coal use in Figure 2-1.

In sum, Figure 2-2 outlines a prompt redirection of effort at the margin that lets us use fossil fuels intelligently to buy the time we need to change over to living on our energy income. The innovations required, both technical and social, compete directly and immediately with the incremental actions that constitute a hard energy path: fluidized beds versus large coal gasification plants and coal-electric stations, efficient cars versus offshore oil, roof insulation versus Arctic gas, cogeneration versus nuclear power. These two directions of development are mutually exclusive: the pattern of commitments of resources and time required for the hard energy path and the pervasive infrastructure that it accretes gradually make the soft path less and less attainable. That is, our two sets of choices compete not only in what they accomplish, but also in what they allow us to contemplate later. They are logistically competitive, institutionally incompatible, and culturally antithetical. Figure 2-1 obscures this constriction of options, for it peers myopically forward, one power station at a time, extrapolating trend into destiny by self-fulfilling prophecy with no end clearly in sight. Figure 2-2, in contrast, works backward from a strategic goal, asks what we must do when in order to get there, and thus reveals the potential for a radically different path that would be invisible to anyone working forward in time by incremental ad–hocracy.

2.7. LOGISTICS AND ECONOMICS

Both the soft and the hard paths bring us, each in its own way and at broadly similar rates, to the era beyond oil and gas. But the rates of internal adaptation meanwhile are different. As we have seen, the soft path relies on smaller, far simpler supply systems entailing vastly shorter development and construction time (see Chapter 5), and on smaller, less sophisticated management systems. Even converting the urban clusters of a whole country to district heating should take only thirty to forty years. Furthermore, the soft path relies mainly on small, standard, easy to make components and on technical resources dispersed in many organizations of diverse sizes and habits; thus everyone can get into the act, unimpeded by centralized bureaucracies, and can compete for a market share through ingenuity and local adaptation. Besides having much lower and more stable operating costs than the hard path, the soft path appears to have a lower *initial* cost because of its technical simplicity, small unit size, very low overheads, scope for mass production, virtual elimination of distribution losses and of interfuel conversion losses, low exposure to escalation and interest, and prompt incremental construction (so that new capacity is built only when and where it is needed).[29]

The actual costs of whole systems, however, are not the same as perceived costs: solar investments are borne by the householder, electric investments by a utility that can float low interest bonds and amortize over thirty years. During the transitional era, we should therefore consider ways to broaden householders' access to capital markets. For example, the utility could finance the solar investment (leaving its execution to the householder's discretion), then be repaid in installments corresponding to the householder's saving. The householder would thus minimize his or her own—and society's—long-term costs. The utility would have to raise several times less capital than it would without such a scheme—for otherwise it would have to build new electric or synthetic gas capacity at even higher cost—and would turn over its money at least twice as quickly, thus retaining an attractive rate of return on capital. The utility would also avoid social obsolescence and use its existing infrastructure. Such incentives have already led several U.S. gas utilities to use such a capital transfer scheme to finance roof insulation in more than 100,000 houses.

Next, the two paths differ even more in risks than in costs. The hard path entails serious environmental risks, many of which are

---

[29] Estimates of the total capital cost of "soft" systems are necessarily less well developed than those for the "hard" systems, but can be calculated well enough, assuming today's technologies, to make a good case that they are cheaper than the hard systems with which they compete as long-term replacements for dwindling fossil fuels. The calculations are given in Chapter Seven, with a summary and comparison in Chapter Eight. The methodology of such cost comparisons is discussed in Chapters One and Three. The general cost advantage of soft over hard technologies is as valid in Northern Europe as in the U.S.

poorly understood and some of which have probably not yet been thought of. Perhaps the most awkward risk is that late in this century, when it is too late to do much about it, we may well find climatic constraints on coal combustion about to become acute in a few more decades: for it now takes us only that long, not centuries or millennia, to approach such outer limits. The soft path, by minimizing all fossil fuel combustion, hedges our bets. Its environmental impacts are relatively small, tractable, and reversible.[30]

The hard path, further, relies on a very few high technologies whose success is by no means assured. The soft path distributes the technical risk among very many diverse low technologies, most of which are already known to work well. They do need sound engineering—a solar collector or heat pump can be worthless if badly designed—but the engineering is of an altogether different and more forgiving order than the hard path requires, and the cost of failure is much lower both in potential consequences and in number of people affected. The soft path also minimizes the economic risks to capital in case of error, accident, or sabotage; the hard path effectively maximizes those risks by relying on vulnerable high technology devices, each costing more than the endowment of Harvard University. Finally, the soft path appears generally more flexible—and thus robust. Its technical diversity, adaptability, and geographic dispersion make it resilient and offer a good prospect of stability under a wide range of conditions, foreseen or not. The hard path, however, is brittle; it must fail, with widespread and serious disruption, if any of its exacting technical and social conditions is not satisfied continuously and indefinitely.

2.8. GEOPOLITICS

The soft path has novel and important international implications. Just as improvements in end-use efficiency can be used at home (via innovative financing and neighborhood self-help schemes) to lessen first the disproportionate burden of energy waste on the poor, so can soft technologies and reduced pressure on oil markets especially benefit the poor abroad. Soft technologies are ideally suited for rural villagers and urban poor alike, directly helping the more than two billion people who have no electric outlet nor anything to plug into it

[30] See Chapter One, note 16, and A.B. Lovins, "Some Limits to Energy Conversion," in D. Meadows, ed., *Alternatives to Growth* (Cambridge, Massachusetts: Ballinger, 1977). The environmental and social impacts of solar technologies have been assessed in a study for the ERDA Solar Division (Chapter One, note 15).

but who need ways to heat, cook, light, and pump. Soft technologies do not carry with them inappropriate cultural patterns or values; they capitalize on poor countries' most abundant resources (including such protein-poor plants as cassava, eminently suited to making fuel alcohols), helping to redress the severe energy imbalance between temperate and tropical regions; they can often be made locally from local materials and do not require a technical elite to maintain them; they resist technological dependence and commercial monopoly; they conform to modern concepts of agriculturally based ecodevelopment from the bottom up, particularly in the rural villages.

Even more crucial, unilateral adoption of a soft energy path by the United States can go a long way to control nuclear proliferation—perhaps to eliminate it entirely (see Chapter Eleven). Many nuclear advocates have missed this point: believing that there is no alternative to nuclear power, they say that if the United States does not export nuclear technology, others will, so the U.S. might as well get the business and try to use it as a lever to slow the inevitable spread of nuclear weapons to nations and subnational groups in other regions. Yet the genie is not wholly out of the bottle yet—thousands of reactors are planned for a few decades hence, tens of thousands thereafter—and the cork sits unnoticed in our hands.

Perhaps the most important opportunity available to us stems from the fact that for at least the next five or ten years, while nuclear dependence and commitments are still reversible, all countries will continue to rely on the United States for the technical, the economic, and especially the *political* support they need to justify their own nuclear programs. Technical and economic dependence is intricate and pervasive; political dependence is far more important, but has been almost ignored, so we do not yet realize the power of the American example in an essentially imitative world where public and private divisions over nuclear policy are already deep and grow deeper daily.

The fact is that in almost all countries the domestic political base to support nuclear power is not solid but shaky. However great their nuclear ambitions, other countries must still borrow that political support from the United States. Few are succeeding. Nuclear expansion is all but halted by grass roots opposition in Japan and the Netherlands; has been severely impeded in West Germany, France, Switzerland, Italy, and Austria; has been slowed and may soon be stopped in Sweden; has been rejected in Norway and (so far) Australia and New Zealand, as well as in several Canadian provinces; faces an uncertain prospect in Denmark and many American states;

has been widely questioned in Britain, Canada and the U.S.S.R.[31];
and has been opposed in Spain, Brazil, India, Thailand, and elsewhere.
Consider the impact of three prompt, clear U.S. statements:

1. The United States will phase out its nuclear power program[32] and
   its support of others' nuclear power programs.
2. The United States will redirect those resources into the tasks of a
   soft energy path and will freely and unconditionally help any
   other interested countries to do the same, seeking to adapt the
   same broad principles to others' needs and to learn from shared
   experience.
3. The United States will start to treat nonproliferation, control of
   civilian fission technology, and strategic arms reduction as inter-
   related parts of the same problem with intertwined solutions.

I believe that such a universal, nondiscriminatory package of policies would be politically irresistible to North and South, East and
West alike. It would offer perhaps our best chance of transcending
the hypocrisy that has stalled arms control: by no longer artificially
divorcing civilian from military nuclear technology, we would recognize officially the real driving forces behind proliferation; and we
would no longer exhort others not to acquire bombs while claiming
that we ourselves feel more secure with bombs than without them.

Nobody can be certain that such a package of policies, going far
beyond a mere moratorium, would work. The question has received
far too little thought, and political judgments differ. My own, based
on the past ten years' residence in the midst of the European nuclear
debate, is that nuclear power could not flourish there if the United
States did not want it to.[33] In giving up the export market that her
own reactor designs have dominated, the U.S. would be demonstrat-
ing a desire for peace, not profit, thus allaying legitimate European

[31] Recent private reports indicate the Soviet scientific community is deeply
split over the wisdom of nuclear expansion. See also *Nucleonics Week*, 13
May 1976, pp. 12–13.

[32] Current overcapacity, capacity under construction, and the potential for
rapid conservation and cogeneration make this a relatively painless course, whether nuclear generation is merely frozen or phased out altogether. For an illustration (the case of California), see R. Doctor et al., *Sierra Club Bulletin*, May
1976, pp. 4ff. I believe the same is true abroad. See Introduction to *Non-Nuclear
Futures* by A.B. Lovins and J.H. Price (Cambridge, Massachusetts: FOE/Ballinger,
1975).

[33] See *Nucleonics Week*, 6 May 1976, p. 7, and I.C. Bupp and J.C. Derian,
"Nuclear Reactor Safety: The Twilight of Probability," *Harv. Bus. School Bull.*,
March-April 1976. Bupp, after a detailed study of European nuclear politics,
shares this assessment.

commercial suspicions. Those who believe such a move would be seized upon gleefully by, say, French exporters are seriously misjudging French nuclear politics. Skeptics, too, have yet to present a more promising alternative—a credible set of technical and political measures for meticulously restricting to peaceful purposes extremely large amounts of bomb materials that, once generated, will persist for the foreseeable lifetime of our species.

I am confident that the United States can still turn off the technology that it originated and deployed. By rebottling that genie (Chapter Eleven) we could all move to energy and foreign policies that our grandchildren can live with. No more important step could be taken toward revitalizing the American dream and making its highest ideals a global reality.

2.9. SOCIOPOLITICS

Perhaps the most profound difference between the soft and hard paths—the difference that ultimately distinguishes them—is their domestic sociopolitical impact. Both paths, like any fifty-year energy path, entail significant social change. But the kinds of social change needed for a hard path are apt to be much less pleasant, less plausible, less compatible with social diversity and personal freedom of choice, and less consistent with traditional values than are the social changes that could make a soft path work.

It is often said that, on the contrary, a soft path must be repressive; and coercive paths to energy conservation and soft technologies can indeed be imagined. But coercion is not necessary and its use would signal a major failure of imagination, given the many policy instruments available to achieve a given technical end. Why use penal legislation to encourage roof insulation when tax incentives and education (leading to the sophisticated public understanding now being achieved in Canada and parts of Europe) will do? Policy tools need not harm lifestyles or liberties if chosen with reasonable sensitivity.

In contrast to the soft path's dependence on pluralistic consumer choice in deploying a myriad of small devices and refinements, the hard path depends on difficult, large-scale projects requiring a major social commitment under centralized management. We have noted in section 2.3 the extraordinary capital intensity of centralized, electrified high technologies. Their similarly heavy demands on other scarce resources—skills, labor, materials, special sites—likewise cannot be met by market allocation, but require compulsory diversion from whatever priorities are backed by the weakest constituencies. Quasi-warpowers legislation to this end has already been seriously proposed.

The hard path, sometimes portrayed as the bastion of free enterprise and free markets, would instead be a world of subsidies, $100 billion bailouts, oligopolies, regulations, nationalization, eminent domain, corporate statism.

Such dirigiste autarchy is the first of many distortions of the political fabric (see Chapter Nine). While soft technologies can match any settlement pattern, their diversity reflecting our own pluralism, centralized energy sources encourage industrial clustering and urbanization. While soft technologies give everyone the costs and benefits of the energy system he or she chooses, centralized systems inequitably allocate benefits to suburbanites and social costs to politically weaker rural agrarians. Siting big energy systems pits central authority against local autonomy in an increasingly divisive and wasteful form of centrifugal politics that is already proving one of the most potent constraints on expansion.

In an electrical world, your lifeline comes not from an understandable neighborhood technology run by people you know who are at your own social level, but rather from an alien, remote, and perhaps humiliatingly uncontrollable technology run by a faraway, bureaucratized, technical elite who have probably never heard of you. Decisions about who shall have how much energy at what price also become centralized—a politically dangerous trend because it divides those who use energy from those who supply and regulate it. Those who do not like the decisions can simply be disconnected.

The scale and complexity of centralized grids not only make them politically inaccessible to the poor and weak, but also increase the likelihood and size of malfunctions, mistakes, and deliberate disruptions. A small fault or a few discontented people become able to turn off a country. Even a single rifleman can probably black out a typical city instantaneously. Societies may therefore be tempted to discourage disruption through stringent controls akin to a garrison state. In times of social stress, when grids become a likely target for dissidents, the sector may be paramilitarized and further isolated from grassroots politics.

If the technology used, like nuclear power, is subject to technical surprises and unique psychological handicaps, prudence or public clamor may require generic shutdowns in case of an unexpected type of malfunction: one may have to choose between turning off a country and persisting in potentially unsafe operation. Indeed, though many in the $100 billion quasi-civilian nuclear industry agree that it could be politically destroyed if a major accident occurred soon, few have considered the economic or political implications of putting at risk such a large fraction of societal capital. How far would

governments go to protect against a threat—even a purely political
threat—a basket full of such delicate, costly, and essential eggs? Already in individual nuclear plants, the cost of a shutdown—often
many dollars a second—weighs heavily, perhaps too heavily, in operating and safety decisions.

Any demanding high technology tends to develop influential and
dedicated constituencies of those who link its commercial success
with both the public welfare and their own. Such sincerely held
beliefs, peer pressures, and the harsh demands that the work itself
places on time and energy all tend to discourage such people from
acquiring a similarly thorough knowledge of alternative policies and
the need to discuss them. Moreover, the money and talent invested in
an electrical program tend to give it disproportionate influence in the
counsels of government, often directly through staff swapping between policy- and mission-oriented agencies. This incestuous position,
now well developed in most industrial countries, distorts both social
and energy priorities in a lasting way that resists political remedy.

For all these reasons, if nuclear power were clean, safe, economic,
assured of ample fuel, and socially benign per se, it would still be
unattractive because of the political implications of the kind of
energy economy it would lock us into. But fission technology also
has unique sociopolitical side effects arising from the impact of
human fallibility and malice on the persistently toxic and explosive
materials in the fuel cycle. For example, discouraging nuclear violence and coercion requires some abrogation of civil liberties[34] ;
guarding long-lived wastes against geological or social contingencies
implies some form of hierarchical social rigidity or homogeneity to
insulate the technological priesthood from social turbulence; and
making political decisions about nuclear hazards that are compulsory,
remote from social experience, disputed, unknown, or unknowable
may tempt governments to bypass democratic decision in favor of
elitist technocracy.[35]

Even now, the inability of our political institutions to cope with
nuclear hazard is straining both their competence and their perceived
legitimacy. There is no scientific basis for calculating the likelihood
or the maximum long-term effects of nuclear mishaps, or for guaran-
teeing that those effects will not exceed a particular level; we know

[34] R. Ayres, *Harvard Civil Rights—Civil Liberties Law Review 10*:369-443
(1975); J.H. Barton, "Intensified Nuclear Safeguards and Civil Liberties," report
to USNRC, Stanford Law School, 21 October 1975; R. Grove-White and M.
Flood, "Nuclear Prospects: A comment on the Individual, the State and Nuclear
Power" (London: Friends of the Earth Ltd./Council for the Protection of Rural
England/National Council for Civil Liberties, 27 October 1976).

[35] H.P. Green, *George Washington Law Review 43*:791-807 (March 1975).

only that all precautions are, for fundamental reasons, inherently imperfect in essentially unknown degree. Reducing that imperfection
would require much social engineering whose success would be speculative. Technical success in reducing the hazards would not reduce,
and might enhance, the need for such social engineering. The most
attractive political feature of soft technologies and conservation—the
alternatives that will let us avoid these decisions and their high political costs—may be that, like motherhood, everyone is in favor of them.

2.10. SOME DEEPER ISSUES

Civilization in the United States, according to some, would be incon-
ceivable if people used only, say, half as much electricity as now. But
that is what they did use in 1963, when they were at least half as
civilized as now. What would life be like at the per capita levels of
primary energy that Americans had in 1910 (about the present
British level) but with doubled efficiency of energy use and with
the important but not very energy-intensive amenities that people
lacked in 1910, such as telecommunications and modern medicine?
Could it not be at least as agreeable as life today? Since the energy
needed today to produce a unit of GNP varies more than one hun-
dred-fold depending on what good or service is being produced, and
since GNP in turn hardly measures social welfare, why must energy
and welfare march forever in lockstep? Such questions today can
be neither answered nor ignored.

Underlying energy choices are real but tacit choices of personal
values (see Chapter Ten). Those that make a high energy society
work are all too apparent. Those that could sustain lifestyles of elegant frugality are not new; they are in the attic and could be dusted
off and recycled. Such values as thrift, simplicity, diversity, neighbor-
liness, humility, and craftsmanship—perhaps most closely preserved
in politically conservative communities—are already, as we see from
the ballot box and the census, embodied in a substantial social move-
ment, camouflaged by its very pervasiveness. Offered the choice freely and equitably, many people would choose, as Herman Daly puts
it, "growth in things that really count rather than in things that are
merely countable": choose not to transform, in Duane Elgin's
phrase, "a rational concern for material well-being into an obsessive
concern for unconscionable levels of material consumption."

Indeed, we are learning that many of the things we had taken to
be the benefits of affluence are really remedial costs, incurred in the
pursuit of benefits that might be obtainable in other ways without
those costs. Thus much of our prized personal mobility is really in-

voluntary traffic made necessary by the settlement patterns that cars create. Is that traffic a cost or a benefit?

Pricked by such doubts, our inflated craving for consumer ephemerals is giving way to a search for both personal and public purpose, to reexamination of the legitimacy of the industrial ethic. In the new age of scarcity, our ingenious strivings to substitute abstract (therefore limitless) wants for concrete (therefore reasonably bounded) needs no longer seem so virtuous. But where we used to accept unquestioningly the facile, and often self-serving, argument that traditional economic growth is essential for distributional equity, new moral and humane stirrings now are nudging us. We can now ask whether we are not already so wealthy that further growth, far from being essential to addressing our equity problems, is instead an excuse not to mobilize the compassion and commitment that could solve the same problems with or without the growth.

Finally, as national purpose and trust in institutions diminish, governments, striving to halt the drift, seek ever more outward control. We are becoming more uneasily aware of the nascent risk of what a Stanford Research Institute group, quoting Bertram Gross, has called ". . . 'friendly fascism'—a managed society which rules by a faceless and widely dispersed complex of warfare-welfare-industrial-communications-police bureaucracies with a technocratic ideology." In the sphere of politics as of personal values, could many strands of observable social change be converging on a profound cultural transformation whose implications we can only vaguely sense: one in which energy policy, as an integrating principle, could be catalytic?[36]

It is not my purpose here to resolve such questions—only to stress their relevance. Though fuzzy and unscientific, they are the beginning and end of any energy policy. Making values explicit is essential to preserving a society in which diversity of values can flourish: an end that a soft energy path seems better suited to serve (see Chapter Nine).

Some people suppose that a soft energy path entails mainly social problems, a hard path mainly technical problems, so that since in the past we have been better at solving the technical problems, that is the kind we should prefer to incur now. But the hard path, too, involves difficult social problems. We can no longer escape them; we must choose which kinds of social problems we want. The most important, difficult, and neglected questions of energy strategy are not mainly technical or economic but rather social and ethical. They will

[36] W.W. Harman, *An Incomplete Guide to the Future* (Palo Alto, California: Stanford Alumni Association, 1976).

pose a supreme challenge to the adaptability of democratic institutions and to the vitality of our spiritual life.

2.11. EXCLUSIVITY

These choices may seem abstract, but they are sharp, imminent, and practical. We stand at a crossroads: without decisive action our options will slip away. Delay in energy conservation lets wasteful use run on so far that the logistical problems of catching up become insuperable. Delay in widely deploying diverse soft technologies pushes them so far into the future that any credible fossil fuel bridge to them has been burned: they must be well under way before the worst part of the oil and gas decline. Delay in building the fossil fuel bridge makes it too tenuous: what the sophisticated coal technologies can give us, in particular, will no longer mesh with our pattern of transitional needs as oil and gas dwindle.

Yet these kinds of delay are exactly what we can expect if we continue to devote so much money, time, skill, fuel, and political will to the hard technologies that are so demanding of them. Enterprises like nuclear power are not only unnecessary but a positive encumbrance, for they prevent us, through logistical competition and through cultural and institutional incompatibility, from pursuing the tasks of a soft path at a high enough priority to make them work together properly. A hard path can make the attainment of a soft path prohibitively difficult in three ways: by starving its components into garbled and incoherent fragments; by changing social values and perceptions in a way that makes the innovations of a soft path more painful to envisage; and by evolving institutions, policy actions, and political commitments in a way that inhibits those same innovations. Though soft and hard paths are not *technically* incompatible—reactors and solar collectors could in principle coexist—the two paths are antagonistic in other and more important ways that, though qualitative and judgmental, are real and unavoidable. As nations, therefore, we must choose one path before they diverge much further. Indeed, one of the infinite variations on a soft path seems inevitable, either smoothly by choice now or disruptively by necessity later; and I fear that if we do not soon make the choice, growing tensions between rich and poor countries may destroy the conditions that now make smooth attainment of a soft path possible.

These conditions will not be repeated. Some people think we can use oil and gas to bridge to a coal and fission economy, then use that later, if we wish, to bridge to similarly costly technologies in the hazy future. But what if the bridge we are now on is the last one?

Our past major transitions in energy supply were smooth because we subsidized them with cheap fossil fuels. Now our new energy supplies are ten or a hundred times more capital-intensive and will stay that way. If our future capital is generated by economic activity fueled by synthetic gas at $30 a barrel equivalent, nuclear electricity at $60–120 a barrel equivalent, and the like, and if the energy sector itself requires much of that capital just to maintain itself, will capital still be as cheap and plentiful as it is now, or will we have fallen into a "capital trap"? Wherever we make our present transition to, once we arrive we may be stuck there for a long time. Thus if neither the soft nor the hard path were preferable on cost or other grounds, we would still be wise to use our remaining cheap fossil fuels—sparingly—to finance a transition as nearly as possible straight to our ultimate energy income sources. We shall not have another chance to get there.

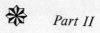 *Part II*

Numbers

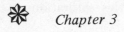 *Chapter 3*

Methods of Exploring the Energy Future

3.1. EXPANSIO AD ABSURDUM VS. PRECISION GUESSWORK

"It is difficult," said Niels Bohr, "to make predictions, especially about the future." But this difficulty has never inhibited those eager to forecast or predict[1] the energy future. Indeed, energy forecasters often claim, or at least imply, such precision that, as Dr. Meghnad Desai once astutely remarked, we appear to know much more about the future than we do about the present or the past.

The methods common to such projections, already criticized by Daly[2] in Chapter One, assume that the future is essentially the past writ large. Quantitative assumptions, such as rates of growth in primary energy demand or in GNP,[3] are extrapolated by so many percent per year—i.e., inherently exponentially—changing according

[1] For the distinction, see H. Daly, *Am. Inst. Plan. J.*, pp. 4-15 (January 1976).

[2] See also H. Daly, ed., *Toward a Steady-State Economy* (San Francisco: W.H. Freeman, 1973).

[3] As was mentioned on p. 37, Chester Cooper and Alvin Weinberg of the Institute for Energy Analysis (Oak Ridge) have argued persuasively ("Economic and Environmental Implications of a U.S. Nuclear Moratorium 1985-2010") that the projections of demographic and economic growth on which all past U.S. energy projections rest are much too high—a correction that, strictly speaking, has not been properly incorporated into Figure 2-2. Of course nobody knows what the future will bring; but if Cooper and Weinberg are wrong, it will not matter as much as one might expect. This is because improvements in energy efficiency occur most rapidly at the margin. The more new houses, machines, etc., that are built, the faster the old, less energy-efficient stocks can be diluted or replaced. Thus people who postulate rapid population and economic growth *and* correspondingly rapid energy growth are trying to have it both ways.

to historically observed patterns (elasticities, saturations, etc.) or not at all.

Such models have trouble adapting to a world in which, for example, real electricity prices are rapidly rising rather than slowly falling as they used to.[4] More generally, such models have a certain inflexibility[5] that tends to lock us into a single narrow vision of lifestyles and development patterns. Extrapolations have fixed structure and no limits, whereas real societies and their objectives evolve structurally over decades and react to limits. Extrapolations have constants, but reality only has slow variables. In a finite world, unlike that of extrapolations, the key questions are those of choice and value within interactive, consecutively approached constraints. Linear programming models, commonly used to test constraints and choices, bear little relation to real decisions, and, if they include enough variables and constraints to be interesting, are likely to be severely unstable, veering between extremes in a way that real societies do not.

Extrapolations assume essentially a surprise-free future even when written by and for people who spend their working lives coping with surprises such as those of late 1973. Formal energy models can function only if stripped of surprises, but then they can say nothing useful about a world in which discontinuities and singularities matter more than the fragments of secular trend in between. Worst, extrapolations are remote from real policy questions. Decisionmakers are seldom called upon to fix a growth rate at x or y percent per year, but rather to apply pragmatic judgment to intricate decisions of detail. If the modeler assumes a single value, or only a few values, for each basic variable, the model may elicit a "my number is better than your number" response; if many values, the model unhelpfully offers policymakers enough options to keep them undecided indefinitely, so they will simply ignore it. What it does not offer them is a sense of the goals and criteria that would make some numbers preferable to others.[6]

[4] L.D. Chapman et al., "Electricity Demand: Project Independence and the Clean Air Act" (FPC Task Force Report, Oak Ridge, Tennessee: Oak Ridge National Laboratory, November 1975). Von Hippel and Williams show (*Bull. Atom. Scient. 32*:10, 18ff. [December 1976]) that the 1975 average price of U.S. residential electricity would have to more than quadruple to return to the constant dollar level it was at in 1940.

[5] B. Fritsch, "From Classical Economics to Global Political Economy," Arbeitspapier 76/6, Institut für Wirtschaftsforschung, Eidgenössliche Technische Hochschule Zürich (August 1976).

[6] Just as the model does not illuminate what should motivate the decisionmaker, so it is seldom clear what should, or does, motivate the modeler to make such models. One makes a model, presumably, because some system is too complex for its behavior to be apprehended intuitively. The same, unfortunately, is then bound to be true of the model, and one will never know whether to believe it or not, nor how far it can be used for guidance, since one does not understand it and has no way to validate it.

Having diagnosed these defects,[7] Monte Canfield, Jr., and colleagues in the Energy Policy Project of the Ford Foundation sought to explore "the enormous gap between the unavoidable and the miraculous"[8] through several scenarios that are policy-oriented, descriptive (not prescriptive), exploratory, and adaptive. They lie "somewhere between a forecast and a fantasy."[9] They are defined not by exogenous numerical assumptions but rather by policy goals and mental models presumed to prevail among governments and their constituencies. Therein lies the great utility of the scenarios, for a reader can identify the one that most nearly reflects his or her own frame of mind, note its consequences, and so gain a fuller understanding of the consequences of the practical choices made every day. Scenarios are also easier than extrapolations to test for resilience[10] in the face of surprises.

This scenario technique[11] is valuable and should be extended. Extrapolation, too, if in skilled hands, has its uses in short-term forward planning over a few years (especially when bolstered by multivariate sensitivity analysis) and has rightly had much talent devoted to its refinement. But both techniques must be supplemented by a third method which highlights what can be done in the long term by recommitting resources in the short term. This method is the broad-brush "working backwards" policy exploration exemplified by Chapter Two and by the Canadian[12], Danish[13], and Japanese[14] studies it cites.

Such an exercise should use a time frame long enough to make

---

[7] M. Canfield, Jr., "Normative Analysis of Alternative Futures (A Methodology Using Scenarios for Policy Analysis)" (Occasional Paper, Aspen [Colorado] Institute for Humanistic Studies, March 1974).

[8] J. Steinhart et al., "A Low Energy Scenario for the United States 1975–2050" (Madison: Institute for Environmental Studies, University of Wisconsin, December 1976), discussion draft.

[9] *Ibid.*

[10] C.S. Holling, "Myths of Ecology and Energy" (Vancouver: Institute of Resource Ecology, University of British Columbia, October 1976) (to be published in 1977 by Oak Ridge Associated Universities in the proceedings of the ORAU symposium *Future Strategies for Energy Development*); H.R. Grümm, ed., *Analysis and Computation of Equilibrium Regions of Stability*, CP-75-8 (Laxenburg, Austria: IIASA, 1975).

[11] Unfortunately the word "scenario" has lately been robbed of specificity by those unfamiliar with its original use in the film industry. Properly, it refers to a description of how future events unfold, described chronologically and at least qualitatively in sufficiently vivid detail that readers can readily imagine themselves participating in the events it describes. People who find it useful to have a word for this concept should defend it from those who have forgotten how to say "projection," "forecast," "plan," "program," "prediction," "sketch," "outcome," "future," "prospect," "trend," "proposal," "development," etc.

[12] See Chapter Two, note 18.

[13] See Chapter One, notes 18 and 25.

[14] See Chapter One, note 18.

major changes in supply and to turn over major capital stocks[15] : Chapter Two, typically, uses fifty years, about the lifetime of a major energy facility ordered today. The method does not pretend to achieve econometric precision (which is often spurious anyway[16]). Nor does it try to simulate markets and prices decades ahead. It instead estimates main terms in the spirit of experimental physics, striving not for elaborate sophistication but for transparent simplicity.[17] As noted in Chapter Two, working backward, by turning divergences into convergences, provides insights unavailable to anyone working forward. One can then work forward, or both ways at once, to devise transitional tactics, iterating as necessary.

Another advantage of the semiquantitative approach to energy futures is that it emphasizes the intimate relationships between energy policy and every other kind of policy. Energy studies done in isolation are unreal and dangerous. The energy problem is intimately related to all the great issues of our day, and is arguably only a symptom of deeper social disorders that the energy strategist cannot safely ignore. Thus when Chapter Two suggests that "The most important, difficult, and neglected questions of energy strategy are not mainly technical or economic but rather social and ethical," it is not referring only to social and ethical *energy* issues. Caldwell sums it up: "The energy crisis is a natural product of the sociotechnical system that it now threatens."[18]

3.2. THE ROLE OF ECONOMICS

Chapters Six through Eight will estimate and compare the capital costs of various energy options. These estimates, and the kinds of conclusions that may be drawn from them, deserve several caveats growing out of the inherent limitations of economics in long-term policy exploration.

The cost calculations presented are approximate, illustrative, even naive. Their style will be more familiar to a physicist than to an

[15] Typical turnover times are of the order of ten years for cars, twenty for industrial plant, and fifty for buildings. These numbers have probably been decreasing steeply in recent years—which, if true, would have important implications for future capital requirements. In a conserver society, physical capital would last much longer.

[16] "It is the mark of the educated man and a proof of his culture that in every subject he looks for only so much precision as its nature permits (or its solution requires)."—Aristotle, *Nichomachean Ethics*, Book I, Chapter 3, paragraph 4, quoted by Gareth Price of the Shell Group in London.

[17] See Chapter Two, note 18.

[18] L.K. Caldwell, "Energy and the Structure of Social Institutions," *Human Ecology* 4:1, 35–45 (1976).

economist. They appear adequate to support the arguments based on them, but they need refinement, which readers are earnestly solicited to attempt. Some may object that the data are highly uncertain. This is correct. Solar heating costs, for example, are nearly as uncertain as nuclear power costs. Commercial U.S. reactors have consistently been commissioned at twice their projected real capital cost[19] and, lately, at three or more times their projected real fuel cycle costs.[20] There are several possible lines of defense against such uncertainty, all of which are used in this book: first, to include in the calculations ample conservatisms in the sense least favorable to the argument, for example, by underestimating nuclear costs and overestimating solar costs; second, to make the assumptions and citations explicit; third, to focus attention on the uncertainties and inadequacies of the analysis, and particularly on the most important terms; and fourth, to note any relevant structural trends, for example, the likelihood that nuclear costs will rise and solar costs will fall. These approaches and several stages of adversarial review have helped to ensure that, within the limited precision to which the results are relied upon, they are as likely to be reliable as any others—probably more likely.

This book makes no attempt to predict the future course of energy markets (or rather, fuel bazaars) or of fuel prices, save to suggest, as in Chapter One, that long-term price must rise because it is far below long-term marginal cost. This omission is not due to lack of interest, but rather to the belief that no such exercise can be worth doing. This is essentially because, as Daly points out,

. . . the prices of resources *in situ* are largely arbitrary.[21] All that economic theory can tell us about the price of resources in the ground is that they must fall between historical cost of production and present cost of replacement—*i.e.* approximately between zero and infinity for non-renewables! The competitive market with its short run vision selects in favor of the zero price convention, but any long run cost accounting would favor an approximation to the replacement cost convention. If cost were defined on a replacement philosophy[,] fossil fuels would become very expensive and solar energy would become ⌊universally⌋ economic overnight. The choice between solar and fossil fuels (and soft and hard technology generally) depends largely on our philosophical definitions of cost (historical

[19] I.C. Bupp et al., "Trends in Light Water Reactor Capital Costs in the United States: Causes and Consequences," Center for Policy Alternatives (MIT), CPA 74-8, 18 Dec. 1974; summarized in *Techn. Rev.* 77:2, 15–25 (February 1975).

[20] For a typical case, worked out in current dollars. see J. Harding, "The Deflation of Rancho Seco 2," FOE (124 Spear St., San Francisco, California 94105), 1975. The nearly fivefold discrepancy he cites is not atypical. See also empirical data in *Electr. World*, pp. 43–58, 15 Nov. 1975.

[21] Because not everyone (over all future time) bids for them: N. Georgescu-

vs. replacement; private *vs.* social) and not on some "value-free numbers generated by an impersonal market." Price calculations can help to achieve either the hard or soft energy future in an efficient manner, but cannot help us choose between basic goals. The choice between the hard and soft energy futures is price-determining, not price-determined.[22]

Thus the prevalent philosophy today, treating fossil capital as if it were income, heavily discounts its depletion. Moreover, if there were free markets in energy—none of which is now observable—energy prices would still depend heavily on future geological and technical contingencies that are now unknowable. In fact, such energy markets as we have and shall have for the foreseeable future are not free, but are distorted by a variety of taxes, subsidies, seminegotiable economic rents, and the like, and are never in equilibrium.

This book extracts useful information from energy economics, despite these pervasive uncertainties, by using capital intensity as an *a posteriori* signal rather than market equilibrium as an *a priori* mechanism. Comparing capital intensity is advantageous because:

1. the high absolute capital intensity, and high ratio of capital to fuel cost, of all significant energy supply technologies at the margin means that capital cost will be an increasingly dominant term in life cycle cost, and, to first order, will provide the long-term signals that one would otherwise seek from life cycle cost[23] ;
2. capital cost is an engineering number that can be calculated, even for future production, with far more confidence than future fuel

---

Roegen, *Southern Econ. J. 41*:3, 347–81 (January 1976). Some economists who should know better say that it is all right to discount away the interests of future generations in today's resource (or, strictly, negentropy) depletion, since technological ability increases exponentially with time and will thus compensate. The basis for this facile view turns out to be a traditional ad hoc identification—with no analytical foundation—of technological progress with the exponentially increasing time factor in Cobb-Douglas production functions. That factor is there only to match Cobb-Douglas results to empirical data—from which they have diverged so far that, according to some analysts, roughly 60 percent of the GNP of many industrial countries is accounted for by the fudge factor! In less shameless disciplines than economics, such a result would long ago have led to revision of the clearly unsatisfactory model; but in fact it has led only to a widespread faith in technology that is often an inverse power function of firsthand knowledge of it and of its thermodynamic limitations.

[22] See Chapter One, note 1.

[23] Perhaps we are moving in this direction already in energy pricing, notably in proposals (e.g., D.A. Casper, *Energy Policy* [U.K.] *4*:191–211 [1976]) to apply more widely the "fuse tax" system said to be in use in some European countries. The system levies a charge on the maximum electrical capacity for which a building is wired, in addition to a charge per kW-h(e) used. In short, the dominance of capital costs in marginal energy (especially electricity) economics provides an incentive to separate the charges for energy and for power (rate of use of energy).

prices, for it is relatively resistant to the interfering effects of tax, price regulation, and oligopoly;

3. the dominant uncertainty in capital cost comparisons tends to be constant dollar escalation rates that, for obvious structural reasons, are likely to be related to the complexity, novelty, maturity,[24] technical adventurousness, political acceptability, and lead time of technologies, and thus amenable to generic comparisons of general trends;

4. engineering calculations provide information that is unobtainable in principle from any kind of economic analysis (e.g., no amount of regressions of historical elasticities can tell you that you can build a heat pump);

5. any soft technology, given ordinarily good engineering, will have an operation and maintenance cost lower than the *sum* of fuel cost and operation and maintenance cost for a hard technology, but the hard technology requires fuel whereas the soft technology (except biomass conversion) does not[25]; hence the soft technology will have a lower life cycle cost than the hard technology if the former has a lower (or even a slightly higher[26]) capital cost than the latter. Further, the robustness of such a comparison can be assessed without an exact knowledge of the margin of difference.[27]

Chapter 1.5 argues that since we are obliged to begin committing resources now to the long-term replacement of historically cheap fuels, we must now compare all potential long-term replacement technologies *with each other, not with the cheap fuels,* in order to avoid a serious misallocation of resources. In this sense, the capital cost comparisons in Chapter Eight are formally equivalent to an assumption of long-run marginal cost pricing, an approach that most

[24] In the sense of B. Commoner, "The Energy Crisis: The Solar Solution" ("Sharing the Sun" Conference, International Solar Energy Society, Winnipeg, 16 August 1976).

[25] Thus, in principle, if income energy were sold rather than homemade, the charge would be entirely for capacity and the energy would be free. (For administrative convenience a small annuity for maintenance might be included in the capacity charge.) This would be a profound change in how we think of energy prices: even the phrase "energy prices" would have to change. In effect, this would be a limiting case of a fuse tax (see note 23, *supra*) in which the power charge is substantial but the energy charge is zero.

[26] A convenient tool here is to express lifetime fuel costs as a present value to be added to whole system capital cost. For example, at a 10 percent/y discount rate—unrealistically high in real terms—a forty-year stream of fuel costs (see note 20, *supra*) for a LWR ordered today has a present value about 42 percent of the reactor's capital cost, heroically assuming that real fuel costs will remain constant. On traditional views, the ratio of present-valued fuel cost to capital cost should be lower for nuclear power than for any other hard technology.

[27] See pp. 135–136 *infra* for an example.

sophisticated energy economists would advocate. There is, however, a subtle difference: such marginal pricing only gives us signals for delivering a fuel, not for performing a function, whereas many soft technologies (such as solar heating) perform the function directly without incurring the costs and losses of further conversion in additional end-use devices. Thus a mere comparison of whole system capital costs must be complemented by an assessment of how efficiently the delivered energy can perform the desired end-use function when converted in the appropriate end-use device (whose capital cost must, of course, be included). Chapter Eight works out one important example (low temperature heat, the dominant term in end-use energy needs), though it is only one example of many.

This difference between delivered fuel and delivered function is seldom reflected in current decisions. This is partly due to a lack of end-use orientation. We are so used to comparing the prices of fuels that few people appreciate that a joule of delivered solar heat is worth—because it replaces—about 1.3 to 1.7 joules of delivered fossil fuel, since the fuel could be burned in a furnace at a First Law efficiency of only, say, 1/1.3 to 1/1.7.[28] The ability of many soft technologies to deliver end-use function directly rather than incurring further costs and losses to convert fuel to function is not reflected in primary energy graphs such as Figure 2-2, but must always be borne in mind, as Figure 2-3 emphasizes.

Another reason that calculations in the style of Chapters Six through Eight may seem novel is institutional. Though a few analysts are starting to look at the whole system cost of performing an end-use function, no institution today does so. The only institutions that perceive whole system costs of delivering a fuel (not a function) are, in general, the vertically integrated oil majors, because they own all the pieces; but their responsibility stops with the delivered barrel and does not traditionally extend to the use made of it. Likewise, utilities are not ordinarily concerned with end-use function, but only seek the cheapest source of large blocks of busbar electricity to feed into a grid.

However well economic comparisons are done, serious questions can be raised about their relevance today. Even Keynes admonished us not to "overstate the importance of the economic problem, or

---

[28] Technical fixes could improve this to 1/0.9 or 1/0.95. Even larger improvements can in principle be achieved by using fuels, at higher Second Law efficiency, to drive a heat pump. Such systems are being developed in Europe. For example, Federal Republic of Germany energy R&D projects #ET5164A–ET5167A commit $4 million over 1975–1978 for Ruhrgas AG, AUDI-NSU, and VW AG to convert Otto and Diesel cycle car engines to burn natural gas to run air to water heat pumps and use the waste heat for boosting. Heat pumps driven by fuel cells are another possibility.

sacrifice to its supposed necessities other matters of greater and more permanent importance."[29] The noneconomic aspects of, say, nuclear power are vastly more important, both theoretically and politically, than its disputed and probably unknowable economic properties.[30]

Moreover, different people have different objective functions, and this is not a mere theoretical nicety but the essence of politics in a democratic republic. Chapter 1.2 suggested some cogent reasons why what looks like cheap energy may really be dearly bought everywhere else in the economy. Paul Ehrlich, taking a slightly different line, has argued[31] that plentiful energy can be considered less a benefit than a cost because of the very expensive mischief that it lets us do (or that its conversion unavoidably does) to essential life support systems, the cost of whose loss is effectively infinite because we do not know how to replace them.

Moreover, though sophisticated managers realize that they should base private investment decisions on sensitivity to altered circumstances rather than on tiny marginal cost differences in the base case,[32] it is not clear that even enormous marginal cost differences are important in public welfare economics. As Alan Poole cogently argues,

> The ultimate condemnation of a project is that it is "uneconomic," which is to say it costs 5 mills per kW-h(e) more than another option. By comparison, human labor in a dirt poor pre-industrial society costs roughly 2000 mills/.5 kW-h(e), 4000 mills per kW-h(e). In this light I find it odd that so little work has been done on the long-term importance of the cost of energy. Does it *really* matter if, say, a solar option costs twice as much as a nuclear option? In the conventional wisdom such a price discrepancy would dismiss the solar option *automatically* even if we were rather concerned about the long-term implications of nuclear power. A pretty good hand-waving case can be made that the effect on economic growth is not very important. A number of European countries appear to have had real energy costs substantially higher than ours[,] yet their per capita GNP is comparable.[33]

[29] Quoted by E.F. Schumacher in *Small is Beautiful* (New York: Harper & Row, 1973).

[30] The economics of nuclear power depends upon some fifty-odd main variables, of which few are known to within a factor of two, all are disputed, and none is orthogonal (independent of the others).

[31] P.R. Ehrlich, "An Ecologist's Perspective on Nuclear Power," Federation of American Scientists *Public Interest Report 28*:5/6, 3–7 (May/June 1975).

[32] The role of base case "reference projections" is often misunderstood. Their purpose is to show what will *not* happen and thus to illuminate why not— *i.e.*, what will fall apart first.

[33] A.D. Poole, "Questions on Long-Term Energy Trajectories" (notes from a talk delivered to the ORAU summer faculty institute [July 1976], IEA, Oak Ridge, Tennessee), typescript.

This is not only to argue, as Alan Kneese does,[34] that the usual cost-benefit or cost-cost comparisons break down[35] when some costs are not quantifiable (and may hence seem less real than others) or are transferred to other times and places (which makes the theoretical basis of cost-benefit comparison invalid); it is to argue, more fundamentally, and within the narrowest of economic paradigms, that since human labor is three orders of magnitude more expensive than inanimate energy, very substantial increases in the comparatively negligible price of the latter may be of no great moment.

[34] A.V. Kneese, "The Faustian Bargain," *Resources 44*:1 (September 1973), Resources for the Future, Washington, D.C.

[35] A.B. Lovins, "Cost-Risk-Benefit Assessments in Energy Policy," *Geo. Wash. L. Rev.*, in press (August 1977).

Chapter 4

Energy Quality

4.1. CALCULATIONS

No definitive, detailed survey of the thermodynamic structure of end-use energy has been done in any country. This
chapter will take a first cut at the problem, first for the U.S. and then,
necessarily in less detail, for some other industrial countries. The data
base used for the U.S. is the Ross and Williams revision[1] for 1973 of
the somewhat problematical, but adequate, SRI data base[2] for 1968.
The 1973 version is shown in Table 4-1.

The end uses shown in Table 4-1 can be conservatively assigned to
broad thermodynamic categories by inspection, except for the large
terms representing industrial process heat (both direct and steam).
The approximate temperature spectrum of these terms, roughly a
fourth of all U.S. primary energy, must be estimated before we can
proceed.

Before 1975, the only attempt to construct a modestly detailed
spectrum or histogram for U.S. process heat appears to be an unpublished study for Westinghouse that gives five temperature categories
for fifteen process sectors[3] in 1972. The conventions and methodol-

---

[1] M. Ross & R.H. Williams, "Assessing the Potential for Fuel Conservation"
(Buffalo: Institute for Public Policy Alternatives, State University of New York,
1975); revised version in *Techn. Rev.*, in press, 1977.

[2] Stanford Research Institute, "Patterns of Energy Consumption in the
United States," PB-212 776 (Springfield, Virginia: NTIS, 1972).

[3] Food and kindred products; textile mill products; apparel and other textile
products; lumber and wood products; paper and allied products; chemical and
allied products; petroleum and coal products; rubber and plastic products; stone,

ogies used are not available. The "fossil fuel uses" (in units of 10^{12} BTU) are shown in summary form in Table 4-2. The figures, though rough, do suggest that it would be useful to confirm how much heat is used in industry at low and moderate temperatures.

Two studies nearing completion for the ERDA Solar Division[4] at this writing are exploring the economic and engineering parameters of solar process heat, and in the process are developing for the first time a detailed temperature spectrum for U.S. industrial process heat. The data currently available are incomplete, preliminary, subject to revision, but nonetheless very interesting, and probably accurate enough for our purposes here.

The unique methodological feature of the ITC study[5] is that

... the survey was performed from the point of view of process requirements rather than the point of view of current methods of using heat. Thus, the temperature of major interest for a particular application was the required temperature of the process material rather than the temperature at which the heat is currently provided. Currently much heat of high thermodynamic availability is wasted because it is used for a low-temperature application which could readily be satisfied with low-temperature heat. Fuels with the capability of a flame temperature of 2000+°F are burned with the ultimate objective of making hot water, for example. A solar process heat system should be designed to satisfy the needs of a process and not merely to substitute for the current method of providing heat.[6]

A further methodological departure of the ITC study is that it examines not only the terminal temperatures desired, but also the corresponding spectrum of temperatures required if the preheat of the process material from ambient temperature is taken into account. Failure to do this would exaggerate the amount of high temperature heat required.

Clearly neither of these novel approaches can rely on current statistical data. The ITC survey therefore selects, from the more than 450 four-digit SIC [Standard Industrial Classification] Groups in

clay, and glass products; primary metal industries; fabricated metal products; machinery, except electrical; electrical equipment and supplies; transportation equipment; natural gas processing.

[4] The project manager is Mr. William Cherry. The two contractors, working independently at ERDA's request, are InterTechnology Corp. (Warrenton, Virginia) and Battelle Columbus Laboratories. Mr. Malcolm Fraser of ITC and Dr. Elton Hall of BCL have generously sent me (October 1976) the early data used here. Both studies are to be published in 1977 through NTIS.

[5] *Ibid.* G.M. Reister, UCRL-51747, Lawrence Livermore Laboratory, February 1975 cites a much higher proportion of low-temperature process heat.

[6] M. Fraser, "Survey of the Applications of Solar Thermal Energy to Industrial Process Heat" ("Sharing the Sun" Conference, International Solar Energy Society, Winnipeg, 15-20 August 1976).

Table 4–1. 1973 U.S. Primary Energy Consumption by End Use (10^{15} BTU/y)

Sector	Direct Fuel	Electricity	Total Fuel
Residential total	7.89	1.97	14.07
space heat	6.16	0.32	7.19
water heat	1.26	0.28	2.13
air conditioning		0.32	1.00
refrigeration		0.38	1.18
cooking	0.38	0.08	0.63
lighting		0.26	0.82
clothes drying	0.09	0.08	0.34
other electrical		0.25	0.78
Commercial total	6.65	1.74	12.06
space heat	4.28		4.28
water heat	0.61		0.61
air conditioning	0.35	0.41	1.63
refrigeration		0.28	0.87
cooking	0.15		0.15
lighting		1.05	3.26
asphalt & road oil	1.26		1.26
Industrial total	21.44	2.96	29.65
process steam	10.54		10.54
electricity generation	0.33		0.33
direct heat	6.58	0.18	7.09
electric drive		2.34	6.48
electrolysis		0.34	0.94
other electrical		0.10	0.28
feedstocks	3.99		3.99
Transportation total	18.91	0.02	18.96
automobiles	9.81		9.81
trucks	3.90		3.90
aircraft	1.29		1.29
rail	0.58	0.02	0.63
pipelines	1.81		1.81
ships	0.26		0.26
buses	0.16		0.16
other	1.10		1.10
GRAND TOTAL	54.89	6.69	74.74
(percent of grand total)	(73)	(9)	(100)
GRAND TOTAL EXCLUDING PETROCHEMICAL FEEDSTOCKS, REDUCTANT COKE, ASPHALT, & ROAD OIL	49.64	6.69	69.49
(percent of grand total so modified)	(71)	(10)	(100)

Table 4–2. 1972 U.S. Fossil Fuel Used To Provide
Process Heat (10^{12} BTU/y)

Industry	$<230^{o}F$	$231-400^{o}F$	"$400^{o}F$ & Stm. Dr."	Electrical Generation	Space Heat	Total
process	2,363	2,944	13,641	732	1,063	20,743
other	708	471	590	24	564	2,357
total	3,071	3,415	14,231	756	1,627	23,100
(% of total)	(13)	(15)	(62)	(3)	(7)	(100)

mining and manufacturing, more than one hundred for detailed process analysis. These one hundred-plus sectors include all the industries that used over 5 TW-h(t) of direct fuels in 1972, plus some others required "to obtain a broad-based coverage of industry".[7] The one-hundred-plus industry sample uses about 70 percent of all industrial direct fuel. For each industry, ITC constructs process flowsheets showing process, temperature, and amount of nonelectric heat needed for specific applications, then integrates the results into a temperature spectrum. The preliminary spectrum available at this writing is redrawn in Figure 4–1. It includes some sixty-seven four-digit SIC Groups (and over 170 specific processes) using a total of 7.2×10^{15} BTU/y, or about 48 percent of the estimated total use of process heat in U.S. industry in 1972. ITC believes the final (larger) data base will not yield radically different curves than this interim sample.

In terms of terminal temperature, Figure 4–1 shows that about 5 percent of the heat load is below 100°C and about 24 percent below 177°C (350°F). But if the heat required to preheat from ambient is also considered—for it "should be possible to visualize a method of utilizing preheat [*e.g.*, from solar sources] in essentially every application"[8] —these figures change to about 28 and 42 percent respectively, as shown in the upper curve in Figure 4–1. (These proportions are the basis of the partitioning of U.S. process heat into components above and below 100°C in Table 4–4 and hence in Chapter 2.5.)

The BCL data,[9] which use the same process-oriented approach, appears less detailed and, at this writing, at an earlier stage of development. The preliminary results of process analysis of an incomplete sample (7.87×10^{15} BTU/y), and of a less fine grained and less precise statistical analysis of a larger sample, are shown in Table 4–3. The sample in the right-hand column reportedly includes about 80 percent of all process heat; the rest is in unexamined small industries with unknown temperature distribution.

[7] *Ibid.*
[8] *Ibid.*
[9] *Supra* note 4.

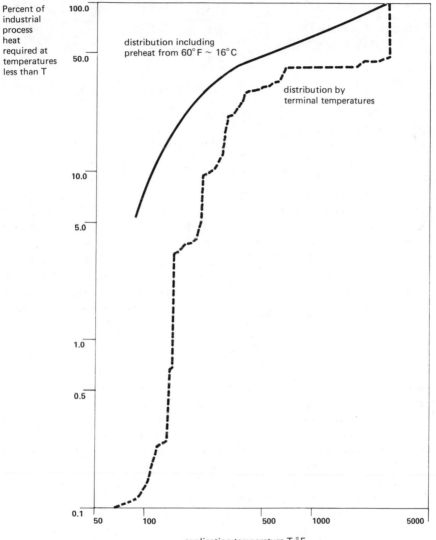

Percent of industrial process heat required at temperatures less than T

distribution including preheat from 60°F ~ 16°C

distribution by terminal temperatures

application temperature T,°F

Figure 4-1. Preliminary ITC Temperature Spectrum of ~ 1972 U.S. Process Heat.

Table 4-3. Preliminary BCL Temperature Spectrum of ~1975 U.S. Process Heat Supplied by Direct Fuel (10^{15} BTU/y)

Category	Process Analysis		Statistical Analysis	
<100°C (<212°F): total	0.27	(2%)	0.26	(3%)
hot water	0.07		0.12	
direct heat or hot air	0.1		0.14	
100-177°C (212-350°F): total	1.34	(17%)	3.25	(32%)
steam	1.2		2.6	
direct heat	0.14		0.65	
>177°C (>350°F): total	6.35	(81%)	6.53	(65%)
steam	0.5		0.56	
direct heat	5.85		5.97	
GRAND TOTAL (rounding errors not corrected)	7.86	(99%)	10.04	(100%)

The BCL data thus show about 2-3 percent of process heat below 100°C and a cumulative total of 19 percent (by process analysis) or 35 percent (by statistical analysis) below 177°C. In view of the inherent uncertainties and limited samples used, the data seem reasonably consistent with the ITC data if the BCL data are assumed to reflect terminal temperatures without regard to preheat from ambient. The greater detail of the ITC data base, however, makes it seem preferable as a basis for this analysis. More useful comparisons will of course be possible once the reports are both published.[10]

The ITC data, then, give us a preliminary basis for completing the thermodynamic classification in Table 4-4, which shows the *approximate* distribution of end use enthalpy according to the type of work to be done (ignoring fuels used as materials and considering only the separable problem of fuels used for their energy content). Note that conservation laws are satisfied: the horizontal sum of the grand total row is 69.49 q, which is the primary energy use (as fuels) shown in Table 4-1. The upper row of percentages at the bottom of Table 4-4 is equivalent to that given in Chapter 2.5. (There the obligatorily electrical component is disaggregated into roughly half industrial electric drive and half other applications.)

A temperature spectrum of process heat has been calculated for Canada by Puttagunta,[11] though the conventions and methodologies used are not stated. Combining these data with others from standard

[10] *Ibid.*

[11] V.R. Puttagunta, "Temperature Distribution of the Energy Consumed as Heat in Canada," AECL-5235, Whiteshell Nuclear Research Establishment, October 1975.

Canadian energy statistics yields the approximate end-use structure summarized here in Table 4-5.[12]

For the Federal Republic of Germany, classifications of end-use energy by general thermodynamic category are available.[13] Unfortunately, classifications that disaggregate heat by temperature range[14] consider only the terminal temperatures of process heat as currently supplied. The two studies available[15] of the temperature spectrum of F.R.G. process heat both use this assumption, and are therefore strongly biased toward high temperatures. No survey of the spectrum of temperatures required for processes, including pre-heat, in the style of the ITC survey[16] is yet available. Table 4-5 therefore assumes that the ITC process heat temperature spectrum for the U.S. is also valid for West Germany, and further assumes that half the obligatorily electrical uses are for industrial drive.

Data for France are unsatisfactory. J.C. Hourcade et al.[17] have estimated French end use in 1973 as 57.3 percent heat (35 percent below 100°C, 5.5 percent 100-200°C, 16.8 percent over 200°C), 25.1 percent mechanical work, and 17.6 percent electrical (10.6 percent industrial, 7 percent domestic and transport). Unfortunately

[12] The structure is shown graphically in A.B. Lovins, *Conserver Society Notes* (Ottawa: Science Council of Canada, May/June 1976), Figure 6.

[13] H. Wilhelms, "Zur Struktur der zukünftigen Energieversorgung," Siemens AG, May 1976, p. 7., Bild 4. The end-use classification given is 24 percent mechanical energy (of which 18 percent is transport), 36 percent process heat, 40 percent space heat.

[14] W. Häfele & W. Sassin, *Energy* 1:147-63 (1976). They state at p. 157 that F.R.G. end use in 1970 was 40 percent industrial (6.4 percent electrical, 24 percent heat over 200°C, 9.5 percent heat below 200°C), 42 percent residential and commercial (4.6 percent electrical, 37.4 percent heat below 200°C, 18 percent transport (0.4 percent electrical, balance heat above 200°C). They do not say how the electricity was used. An unpublished analysis by Kraftwerkunion is very similar: 58 percent heat below 500°C (16 percent industrial, 18 percent commercial, 24 percent household), 17 percent heat over 500°C (all industrial), 10 percent public and commercial transport, 8 percent individual transport, 7 percent other.

[15] Dr. Peter Penczynski has kindly called to my notice two rather detailed process heat spectra for West Germany (in addition to the data given in note 13, *supra*). The spectrum most often cited is by Programmgruppe Systemforschung und Technologische Entwicklung, Kernforschungsanlage Jülich, *Einsatzmöglichkeiten neuer Energiesysteme: I. Bedarfsanalyse und Strom* (Bonn: Bundesministerium für Forschung und Technologie, 1976). It gives at p. 4 both a bimodal temperature spectrum for F.R.G. process heat and its integral, which shows more than half above 1000°C and only about 10 percent below 200°C. A ten-sector analysis by Professor Schäfer of Institut für Energiewirtschaft (Munich) gives a similar result: "Aktuelle Wege zur verbesserten Energieanwendung" (Verein Deutscher Ingenieure, Berichte Nr. 250, 1975), p. 10.

[16] *Supra* note 4.

[17] Unpublished preliminary estimates by J.C. Hourcade, F. Moisan, & J. Zoghbi, Centre International des Recherches sur l'Environnement et le Développement, Paris.

Table 4–4. 1973 U.S. Deliveries of Enthalpy by End Use (10^{15} BTU/y)

Sector	Heating and Cooling		Portable Liquids	Miscellaneous Mechanical Work	Obligatorily Electrical[a]	Lost at Power Stations
	$\Delta T < 100°C$	$\Delta T \geqslant 100°C$				
Residential total	8.72	0.62			0.52	4.21
space heat	6.48					0.71
water heat	1.54					0.59
air conditioning	0.32					0.68
refrigeration	0.38					0.80
cooking		0.46				0.17
lighting					0.26[b]	0.56
clothes drying		0.16			0.01[b]	0.17
other electrical					0.25	0.53
Commercial total	5.93	0.15			1.05	3.67
space heat	4.28					
water heat	0.61					
air conditioning	0.76					
refrigeration	0.28					0.87
cooking		0.15				0.59
lighting					1.05	2.21
Industrial total	4.84	12.46			2.78	5.58[c]
process heat	4.84	12.46				0.33[c]
electric drive					2.34	4.14
electrolysis & other electrical					0.44	0.33[c]
Transportation total			17.10	1.81	0.02	0.03
autos, trucks, buses, aircraft, ships			15.42			
rail			0.58			
other			1.10	1.81	0.02	0.03
GRAND TOTAL	19.49	13.23	17.10	1.81	4.37	13.49
(percent of 56.33 q[d])	(35)	(23)	(30)	(3)	(8)	—
(percent of 69.49 q[e])	(28)	(19)	(25)	(3)	(6)	(19)

a All these applications are assumed to be obligatorily electrical even though some are likely in practice to be readily substitutable, e.g., by gas mantle or solar lighting or compressed air drive.

b Assumed to be the drive (as opposed to heat) requirements.

c The 0.33 q of fossil fuel used to generate electricity at industrial sites is arbitrarily assumed to provide its output to "other electrical"; the exact mix is unknown.

d The total delivered enthalpy shown in Table 4-1 is 49.64 + 6.69 = 56.33 q.

e The gross primary energy shown in Table 4-1, excluding nonenergy uses, is 56.33[d] + 13.49 − 0.33[c] = 69.49 q.

Table 4-5. Approximate Deliveries of Enthalpy by End Use (percentage)

Category	U.S. 1973	Canada ~1973	F.R.G. 1975	France 1975	U.K. ~1975
Heat	58	69	76	64	65
<100°C	35	39	50	37	55
100–200°C	6[a]	19	6	}27	}10
>200°C	17	11	20		
Mechanical work	38	27	21	31	30
vehicles	31	}24	}18	}26	}27
pipelines	3				
industrial electric drive	4	3	3	5	3
Other electrical[b]	4	4	4	5	5

a Includes appropriate share of process heat (see text), all clothes drying, and two-thirds of cooking.

b Includes all lighting, electronics, telecommunications, electrometallurgy, electrochemistry, arc welding, electric drive for public transport and for home appliances, and all other uses of electricity except low grade heating and cooling (under heat) and industrial electric drive (under work).

N.B.: Many of the data are rough and preliminary. See text for detailed reservations. The data for France and the U.K. are especially unsatisfactory, though probably among the best available.

the data, though a good beginning, are not useful in this form, for several reasons. First, the heat spectrum appears to have the same deficiencies as the F.R.G. data. Second, the electrical sector estimate includes many applications that are not appropriate and would ordinarily be covered by fuel but for Electricité de France's aggressive electrification program since the late 1960s. Hourcade estimates[18] that about 10 percent out of the 17.6 percent electrical uses are appropriate, but it is not clear how the remaining 7.6 percent or so should be allocated to other sectors. As a very rough correction, Table 4-5 assumes that the electrical share in the original data is allocated 5 percent to industrial electric drive, 5 percent to other appropriate electrical applications, and 7 percent to heat, split 2 percent to low and 5 percent to high temperatures. These assumptions appear consistent with such statistics as are available, but cannot be considered reliable within better than a few percentage points.

For the United Kingdom, rough estimates from national statistics (including Census of Production data), broadly consistent with estimates by the Energy Research Group of The Open University, are shown in Table 4-5, but might be wrong by five or even ten percentage points. The bias toward low temperature heat is typical for Europe.

Few independent checks are available for any country, and reliable comparative data are few. Data for such countries as Australia, Poland, and Israel are becoming available, though not yet in systematic enough form to present here, while obtaining data for developing countries is a high priority, but good studies are scarce. The first estimates[19] of this kind to be published—by the American Physical Society study group—are consistent with the U.S. column in Table 4-5, which merely reproduces data from Table 4-4.

4.2. INTERPRETATION

Table 4-4 is eloquent and repays close study, but what is it saying? Of course it takes no account of end use effectiveness or efficiency— of how much of a stove's heat, or the primary energy used to produce it, gets into the pot, or of the ability of 0.32 q of electricity

---

[18] *Ibid.*

[19] APS (see Chapter Two, note 11) estimated, for the U.S. in 1968, that end use was about 60 percent heat (28½ percent low temperature, 19 percent process steam and cooking, 12 percent high temperature direct heat), 26 percent vehicles, and 14 percent special applications—rather close to the present 1973 estimate of 58, 31, and 11 percent (including pipelines), respectively. Process heat spectra (see note 4, *supra*) were of course not yet available then.

applied to residential air conditioners to move about 0.61 q of un-wanted heat[20] into the volume of air partly cooled by a neighbor's air conditioner. But the coefficient of performance (COP) attain-able from electrically driven heat pumps such as refrigerators may not be as important as we like to suppose. Heat pumps can be driven by any source of mechanical work (including wind hydraulic drive), and refrigeration, as Robert Williams has pointed out, can be advantageously integrated into the heating and cooling system of a house (even a solar house) rather than restricted to a special plug-in appliance.

Even the undoubted efficiency of electricity for driving motors pales a bit when we recall that mechanical work in fixed devices is only about 7 percent of all U.S. end use; that fossil fuels are also very high grade forms of energy and are often used at very respect-able First Law efficiencies; and that the First Law efficiency of the electric motors studied by Goldstein and Rosenfeld[21] has fallen from 71 to 49 percent for a *single manufacturer* since 1940 (and to 30 percent for another manufacturer). Thus we must not accept too uncritically the facile assumption that electricity is bound to have several times the social or economic effectiveness of, say, fossil fuel. In practice its advantage is highly variable, needs detailed study for each case, and may be significant, negligible, or conceivably even negative. As we know to our cost from fossil fuel end-use technology, the effectiveness of premium forms of energy in practice may diverge markedly from what engineering theory and assumptions of rational behavior lead us to expect.

There is room for infinite argument about the significance of the right-hand column of Table 4-4, which is shown for clarity and com-pleteness. But however one takes account of the low entropy attained by partitioning entropy (and losing enthalpy) at power stations, it is clear that the specialized applications that can take greatest advan-tage of electricity's high quality are but a small fraction of all end uses. Likewise it is clear that high grade fuels like natural or synthetic gas[22] are unnecessary for a large fraction of our heat needs, even in industry[23] —a point made even more emphatically in Table 4-5 for

[20] The average COP of currently installed U.S. air conditioners is given as about 1.9 in several recent studies, e.g., R-1641-NSF (p. 31) and ORNL-NSF-EP-51.

[21] D.B. Goldstein & A.S. Rosenfeld, "Conservation and Peak Power Demand," LBL-4438 (Berkeley, California: Lawrence Berkeley Lab., 1975); see also their "Projecting an Energy Efficient California", LBL-3274, 1975. In a similar vein, Professor Murgatroyd of Imperial College (London) estimates that simple adjust-ments to U.S. domestic refrigerators could save 10 GW(e) peak by 1980.

[22] *Cf.* Chapter 3, note 28.

[23] *Supra* note 6.

countries other than the U.S. The reasons for restricting electricity and premium fuels as nearly as possible to those few end uses that are truly advantageous—cost, First and Second Law efficiency, and sociopolitics—will be considered more fully below, particularly in Chapter 8.2. Meanwhile we need merely note that approaching the energy problem from the viewpoint of end use structure, rather than of the forms of energy most readily made from the most abundant domestic fuels, gives a completely different impression of what kinds of energy we need.[24]

[24] Similar methodological lessons—being demand oriented, matching quality to end-use needs, etc.—are just starting to be learned in other resource problems such as water. A systematic survey of what we have learned the hard way in energy policy should be useful elsewhere.

Chapter 5

Scale

5.1. IS SCALE AN ISSUE?

Since energy is but a means to social ends, and since energy is useful only insofar as it performs specific tasks relevant to those ends, anyone seeking to perform tasks with the least possible energy and trouble should start with an inventory of tasks. Specifically, to design a rational energy system, we need to know the unit scale, type, quality, and degree of geographical clustering of energy needs as a function of space and time. Such an elementary data base does not exist today anywhere in the world. No country even appears to have a usable inventory of the scale of individual end-use energy needs.[1] One must instead fall back on naive order of magnitude observations—simple but instructive.

For example, the heating load of houses is typically measured in 10^3 W; the peak power output of a car is of order 10^5 W; the total input to one of our society's largest integrated industrial facilities—certain primary metals and isotope enrichment plants—is of order 10^9 W with current designs (which might arguably be different were it not for large dams and power grids). With isolated exceptions, mainly pathological cases associated with large (often heavily sub-

[1] ERDA will have to develop such a data base of end use scale and quality for a typical "Middletown, U.S.A." hypothetical town to be constructed on paper as part of a new study of decentralized energy systems, begun in December 1976 with the author's help. The model town is to be tracked through evolution along hard and soft paths as a heuristic device. A complementary study by a contractor is to examine the social implications of comparative decentralization of energy systems.

sidized) dams or with coal complexes, virtually all our end-use systems are probably smaller than about 10^8 W (the inefficient World Trade Center in New York is wired for a peak electrical load of 80 MW), and a great many—probably the overwhelming majority—are clustered in domestic, commercial, and industrial units smaller than 10^6 W. Most of the end-use devices important to our daily lives require 10^{-1} to 10^3 W and are clustered within living or working units requiring 10^3 to 10^5 W. Most production processes of practical interest can be, and long have been, carried out in units of roughly that scale.

Thus it is not obvious, prima facie, that energy must be converted in blocks of order 10^8–10^{10} W. The arguments usually given for such large scale include reduced unit capital cost (typically by a two-thirds-power scaling law), increased reliability through interconnection, sharing of capacity among nonsimultaneous users, centralized delivery of primary fuel, ease of substituting primary fuels without retrofitting many small conversion systems, localization and hence simplified management of residuals and other side effects, ability to use and finance the best high technologies available, ease of attracting and supporting the specialized maintenance cadre that such systems require, and convenience for the end user, who need merely pay for the delivered energy purchased as a service without necessarily becoming involved in the details of its conversion.

These contentions are not devoid of merit. Big systems do have some real advantages—though advantages are often subjective, and one person's benefit can be another's cost. But this chapter will suggest that many of the advantages claimed for large scale may be doubtful, illusory, tautological, or outweighed by less tangible and less quantifiable but perhaps more important disadvantages and diseconomies.

This balance is perhaps least difficult to illustrate in the electrical sector, where, according to Marchetti, "the power of the largest generator since the beginning of the century has doubled every 6.5 years through five orders of magnitude."[2] Häfele points out that since this doubling time is slightly shorter than that of electricity demand (of order ten years), the electrical system must have become steadily more centralized. Large unit scale is intimately linked with centralization, a harder thing to measure but one with many important sociopolitical effects (see Chapter Nine).

Chapter 2.5 has mentioned some of the potential technical and economic advantages of small scale[3] :

[2] C. Marchetti, RR-76-1 (Laxenburg, Austria: IIASA, 1976), p. 222.
[3] See Chapter One, note 14. The Intermediate Technology Development Group (25 Wilton Road, London SW.1, England) of Schumacher, McRobie, and others has proved small-scale concepts in practice, notably in the manufacture of

1. virtual elimination of the capital costs, operation and maintenance costs, and losses of the distribution infrastructure;
2. elimination of direct diseconomies of large scale, such as the increased need for spinning reserve[4] on electrical grids;
3. major reductions in indirect diseconomies of large scale that arise from the long lead times of large systems; and
4. scope for greatly reducing capital cost by mass production if desired.

Let us consider the first three of these points in turn.

5.2. DISTRIBUTION OVERHEADS

Few people appreciate the diseconomies of distributing centrally generated electricity in a country whose load density, averaged over 1965-1971, ranged[5] from 0.61 to 230 mW/m^2 (mean about 30 mW/m^2). The corresponding load density for residential and small light and power consumers ranged[6] from 1.2 to 500 mW/m^2, and the mean was about 15 mW/m^2, a level[7,8] only about 1/12,000 the density of solar input. These low densities are poorly matched to centralized sources. A recent study finds:

> Transmission and distribution costs contribute significantly to the total costs of providing electrical service. In 1974, privately-owned electric utilities in the United States spent about 35% (over $7 billion) of their total capital expenditures for transmission and distribution equipment.

---

bricks, Portland cement, sugar, and egg cartons. That experience, and design studies now being implemented for paper, glass, chemicals, and other products, offer ample evidence that increased efficiency, low overheads, and low transport costs can produce a wide range of products more cheaply in a small than in a large factory. Indeed, *Fortune* (March 1971, p. 76) reports successful "regional, even local" steel plants in the U.S. See the "Diseconomies of Scale" bibliography of the California Office of Appropriate Technology (Box 1677, Sacramento); other OAT topical bibliographies; and the work of Commissioner I.C. Puri, Small Industries Office, Development Commission (Nirman Bhavan, Maulana Azad Road, New Delhi 11) and of Operation Bharani, University of Mysore.

[4] Trying to judge the true cost of prompt spinning reserve leads one into dense definitional thickets, but it is a big number, and someone must pay for it.

[5] M.L. Baughman & D.J. Bottaro, "Electric Power Transmission and Distribution Systems: Costs and Their Allocation," Research Report #6, NSF-RA-N-75-107 (Austin: Center for Energy Studies, University of Texas, July 1975). The discussion of T&D costs appears not to take account of T&D *losses*.

[6] *Ibid.*

[7] This implies that an average 1 GW(e) plant serves a radius of 146 km; the actual average haul length is of order 343 km. In Europe, according to Marchetti (see note 2, *supra*, p. 221), it is only 100 km, even though the grid is interconnected from Sweden to Sicily. The denser European grids generally correspond to somewhat lower grid costs than those given in Table 5-1 for the U.S.: typically in Europe transmission and distribution together cost about the same as generation.

The expenditures for operation and maintenance of this equipment were about $3.0 billion, an amount equal to about 1/2 the total costs of fuel in 1972.

The costs derived from the transmission and distribution (T&D) system have historically comprised about 2/3 the costs of producing and delivering electricity to residential-commercial customers, and over 1/3 the total costs [of] supplying electricity to large industrial customers.[9]

The study concentrates on major terms—high voltage transmission lines, distribution lines, and operation and maintenance costs of the T&D system. These three terms together account for about 80 percent of total T&D costs.[10] When capital costs are converted into m$/kW-h(e) charges at a 13.5 percent/y fixed charge rate, the remarkable results shown in Table 5-1 emerge.

The authors conclude that:

> . . . almost 70% [69.3%] of the costs of power to residential and small light and power customers are related to transmission and distribution. Of this 70%, almost half [49.3% of the total T&D costs] can be attributed to costs of installing transmission and distribution lines, the two items of T&D equipment that exhibited the most significant regional cost variations. For large light and power customers on the other hand, transmission equipment related costs are only 34% of the total cost of power, while generation comprises about 55%. Distribution equipment and operation and maintenance, including billing, comprise the other 11%.[11]

To put it more plainly, for the smaller customers (average load 1.04 kW[e]), who accounted for about 55% of the annual electricity sales in the whole sample, a dollar spent on electricity was allocated approximately 19 percent to transmission equipment, 24 percent to distribution equipment, 21 percent to operation and maintenance of all that equipment (including metering—a small term—and billing), about 6 percent to profit and arithmetic discrepancies in the analysis (owing largely to differential escalation of various components)—and *only 29 percent to electricity.*

This seems an undeniable diseconomy of centralization. Utility ratepayers would doubtless be startled to learn that they are paying

---

[8] Both sets of data refer only to investor-owned utilities in the forty-eight coterminous states. These utilities had about 80 percent of the total U.S. utility business with about 258,000 large customers and 54.2 million small ones—whereas all utilities together had about 363 million of the latter, implying that private utilities have creamed off the highest density (lowest T&D cost) markets.

[9] *Supra* note 5.

[10] *Ibid.*

[11] *Ibid.*

Table 5-1. Average Components of Cost of Electricity
Bought from U.S. Private Utilities, 1972[a]

(¢/kW-h (e))

| | Class of Customer | |
Item	Residential and Commercial	Industrial
Transmission equipment	0.45	0.43
Distribution equipment	0.58	0.06
Operation and maintenance of T & D	0.50	0.08
TOTAL T & D	1.54	0.57
Generation (estimated)	0.69	0.69
DELIVERED TOTAL COST (estimated)	2.23	1.26
AVERAGE REVENUE (to compare)	2.37	1.17

[a]See note 5, *supra*.

2.2 times as much to have the electricity delivered to them as to generate it. For the large customers (average load 177 kW[e]), the mix is less extreme—generation cost is 1.2 times delivery cost—but still odd, and suggests that there are strong economic incentives for centralized utilities to encourage use patterns that are correspondingly centralized. The wide regional variation in these values—the 1972 T&D cost ranged from 1.0 to 2.3 ¢/kW-h(e) for small and from 0.36 to 0.82 ¢/kW-h(e) for large customers—does not alter the startling size of the T&D overheads.[12] It does not appear from the data presented[13] that truly marginal (rather than recent historic) costs would tell a very different story. Nor does it appear that public utilities, about a quarter the total size of the private ones surveyed,[14] would yield results much different. Since over thirty million Americans currently buy their electricity from small public utilities—notably the 1400-odd members of the American Public Power Association—it would be interesting to check, in the light of scale considerations,

[12] Since 1 bbl = 5.8 GJ and 1 kW-h = 3.6 MJ, 1 ¢/kW-h(e) is equivalent on a heat-supplied basis to $16.10/bbl, a handy number to remember. For example, if the whole system capital cost of a delivered kW(e) from a marginal nuclear system is, say, $3496 in 1976 dollars (see Chapter 6.1), then a fixed charge rate of, say, 0.12/y yields an annual capital charge of $419 spread over 8766 h—i.e., an electricity cost of 4.79¢/kW-h(e) or $77/bbl enthalpic equivalent. (We do not need here to divide by capacity factor nor to multiply by a factor representing transmission and distribution costs and losses, since all that is already included in our whole system capital cost. We have, however, charitably assumed that the fuel, operation, and maintenance are free, rather than about 8.7 m$/kW-h(e) (1976 $) [See Harding, Chapter Three, note 20].)

[13] *Supra* note 5.
[14] *Ibid.*

APPA's contention that it supplies electricity at least as cheaply from small, often "outmoded," power stations as a big modern utility could do, and how far this claim is independent of regulatory differences.

Smaller scale can save capital not only by reducing distribution infrastructure but by sharing various kinds of infrastructure with other devices and enterprises. Many classic examples occur in the integration of methane digesters into food systems (see Section 5.4). Consider, for example, the forty-cow Vermont dairy farm studied at Dartmouth College.[15] Anaerobic digestion of the manure would ordinarily be unattractive in Vermont because the cold climate would entail auxiliary heating. But the heat currently rejected to the air by the bulk milk cooler on the same farm—the device that pumps away (promptly, as required by law) the residual heat from the cows—just suffices, if the two systems are integrated, to keep the digester warm enough to work efficiently. The net yield of methane (to say nothing of high quality fertilizer by-product) is then sufficient to supply all the energy needs of six such dairy farms and their inhabitants.[16,17] Such elegantly simple integration is impossible at large scale. For example, West Germany's addiction to GW(e) scale has led to plans for a costly national grid of hot water pipes for district heating, because the heat rejected by each power station is far too great to be used locally and must thus be exported for long distances.[18]

5.3. RELIABILITY AND INDIRECT EFFECTS

Another direct diseconomy of large scale is the increased likelihood and consequence of failure. Likelihood increases because of loss of technical diversity, pushing technologies and materials too hard, scaling up too fast, requiring major long-term commitments to possibly unpromising lines of development, and the like. Consequence increases because it is much more embarrassing to a grid, for example, to lose 1 GW(e) than 100 MW(e), and technically much harder (and more expensive) to provide spinning reserve against the former event than the latter.

[15] "Energy Policy for Vermont," working paper (Hanover, New Hampshire: Dartmouth College, System Dynamics Group, Spring 1976).

[16] *Ibid.*

[17] See J.C. Voorhoeve, *Ceres*, March-April 1976, pp. 48–50, for other examples of integrated systems.

[18] Bundesministerium für Forschung und Technologie, *Einsatzmöglichkeiten neuer Energiesysteme: V. Fernwärme* (and *IV. Fernenergie*) (Bonn, 1975).

It is now well known that large electrical components, notably turbogenerators, often lose in reliability—hence in contributions to grid operating costs, standby capacity costs, grid instability, and lost revenues—what their size gains in unit capital cost. These unforeseen effects are so common that there is mounting evidence[19] that most types of power stations can have lower busbar costs in sizes of order 10^8 W than of 10^9 W. A recent thermal plant sample[20] illustrates this point (Figure 5-1), and Komanoff's nuclear analysis[21] tends to confirm it. The cause is not hard to find: ever-increasing scale encountering technical diminishing returns. Any effort that scales up through five orders of magnitude and keeps doubling every 6.5 years is bound to come unstuck eventually. (At this rate, in the year 2064 a single turbogenerator would have a capacity of 8 TW(e)—the present total world energy conversion rate!)

Dispersed generation near load centers is well known to improve system integration and stability. Traditionally a "dispersion credit" (of order $100/kW(e)) is assigned to local supply, for example from battery banks or fuel cells associated with distribution facilities, because they save roughly the amount of money that would otherwise be needed for transmission facilities. But this credit may be far too low because it does not count improved stability, hence savings on reserve margin. One recent study states:

For one small system, and with assumptions that we believe are realistic, it was found that one kW of dispersed generation was equivalent from the standpoint of reserve requirements to 2.5 kW of central generation. The

[19] W.L. Wilson, *Energy World* 26:2 (April 1976) (Institute of Fuel, London); *Weekly Energy Report* 3, 10:10 (10 March 1975) reporting the FEA avail ability study; R. Mauro, "Statement on Small-Scale Electrical Systems," CEQ Hearings on the ERDA RD&D Plan (Austin), 16 December 1976; R. Knox, *New Scient.* 73:458 (24 February 1977). For a study of component reliability, see H. Procaccia, "Probabilité de défaillance des circuits de refroidissement normaux et des circuits de refroidissement de secours des centrales nucléaires," IAEA-SM-195/13, in *Reliability of Nuclear Power Plants* (Vienna: IAEA, 1975), at pp. 351-72 and discussion p. 392; note particularly Figures 6-9.

[20] Nineteenth Steam Station Survey, *Electr. World*, 15 November 1975, pp. 43-58 (especially the graph on p. 51).

[21] C. Komanoff, "Power Plant Performance" (New York: Council on Economic Priorities, 1976) and addendum thereto (March 1977). Komanoff's multiple regressions are the best that can be done with the data, and he has used all the U.S. data available. It is of course hard to tell whether the unreliability of large reactors is because they are large or because they are new or both, since both are closely correlated. Accordingly, the correlation coefficients that Komanoff obtains, though significant, are not absolutely conclusive. A better case is made, however, with the broader data base of coal-fired plants (for which, in the sense of Figure 5-1, the optimal size appears to have been overshot by a factor of two), and will be able to be made for nuclear plants after a few more years' experience.

Figure 5-1. Average Unavailability and Capital Cost as a Function of Size for Twenty-Nine Recently Built Thermal Power Stations in the U.S.[a]

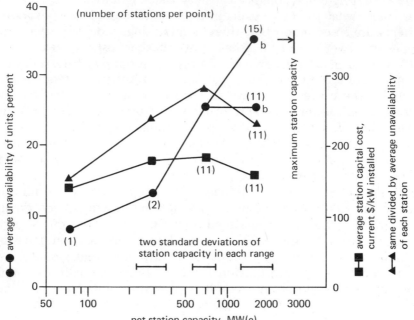

[a]Plotted from tabular data in *Electrical World* (note 20, *supra*), which had plotted the data in a striking graph entitled "Excessive shutdowns slash output of large units" but refused permission to reproduce the graph, on the grounds that the results were invalid owing to an inadequate sample. *Electrical World* had not called this to the attention of its readers, nor observed that Komanoff[21] had obtained strong size-unavailability correlations in virtually the whole population of U.S. nuclear and coal-fired stations. The sample plotted here includes half the U.S. thermal power stations installed in a two-year interval during 1972-4, represents every region except New England, and includes 17-18 coal-fired, 2-4 nuclear, 3 gas-fired, 2 gas- or oil-fired, and 1-2 oil-fired stations.
[b]The lower point is from the tabular data in the Survey[20]; the upper point is from the graph[20], and presumably reflects high unavailability in the four very large stations whose unavailabilities are not shown in the tables. Elsewhere the tables and the *Electrical World* graph are consistent.

reliability of supply within the network was determined by means of an index related to the LOLP [loss of load probability]. The reason why the dispersed device can be so effective is that it protects the load in its vicinity against generation *as well as* transmission and distribution outages.[22]

[22] J. Peschon et al., "Regulatory Approaches for Handling Reliability and Reserve Margin Issues and Impact on Reserve Margin of Dispersed Storage and Generation Systems," testimony to California Energy Resources Conservation and Development Commission (Sacramento), 28 July 1976 (Systems Control,

Studies in 1977 by Edward Kahn (University of California at Berkeley) and by the Wisconsin Public Utility Commission suggest that as a reliable source of bulk marginal power, x GW(e) of medium-sized coal-fired stations would be equivalent to $1.43x$-$1.50x$ GW(e) of large nuclear stations—a ratio that far exceeds the marginal-cost differences on which such investment decisions tend to turn. This effect is due to the lower reserve requirements of the smaller stations, which are unlikely to fail simultaneously. The extreme capital intensity of any electrical system offers ample reason to explore carefully the effects of scale and centralization on system reliability. The classical literature of this subject seems sketchy and unpersuasive and its quantification primitive.[23]

The reliability of different energy systems must be considered symmetrically. Some opponents of soft technologies have suggested, for example, that in order to be as reliable from the user's point of view as electric heat, any solar heating system must be duplicated. But this argument stands reality on its head. The reason electrical grids are designed to such exemplary—and expensive—standards of reliability is that they must be, because so many people depend on them that a failure could be a social catastrophe. If your solar system fails (which, of course, it should not do, as there should not be much to go wrong with it), you can put on a sweater or go next door until it is fixed. But if the electrical grid fails, there is nowhere else to go and not much you can do about it. Indeed, the very scale of the electrical grid makes it unable to discriminate between different users' differing needs for reliability. Everyone, even the person who uses electricity for a water heater and wouldn't even know if it were off for a few hours, must pay for the reliability demanded by hospitals, skyscraper elevators, subways, and traffic lights. A less centralized system could provide for each use only as much reliability as it requires and justifies.

As Dennis Meadows has pointed out, scale and reliability issues can be illuminated by considering the 1976-1977 winter shortages of natural gas in the U.S. Let us ignore for a moment the egregious blunders that so inevitably led to the shortage in the first place, and consider only how its effects were worsened by reliance on a national gas grid. The grid does not make it much easier to allocate supplies—especially since there are bottlenecks in pipeline capacity—but it does

---

Inc., Palo Alto, California). (Emphasis in original.) The calculation is valid for small additions to the grid.

[23] M.L. Telson (*Bell J. Ec.*, Autumn 1975, pp. 679-94) makes a good start on reliability, though not examining scale issues, by showing that the marginal cost of reliability obtained by installing reserve margin in conventional large grids is probably several orders of magnitude larger than its marginal benefits. See also FEA, PB-255 221/4WE (Springfield, Virginia: NTIS, 1976).

guarantee awkward systemic effects. When, for example, gas pressure in a region of the grid drops below a threshold, pilot lights go out and huge (probably nonexistent) armies must be rapidly deployed to go into millions of buildings and turn off valves. Recovering from such a debacle can take weeks. Suppose, as a thought experiment, that exactly the same overstressed reserves of gas were instead being distributed in a decentralized way, for example, in bottles (as in Israel). Then systemic effects would not occur, whatever power the State had to allocate supplies could be used anyway, and people would not all run out of gas at once over whole regions. In the spirit of the previous paragraph, someone whose bottle ran out could go next door. It is not only a lesser dependence on natural gas but also a less centralized system of using gas that has largely protected New England from the distress felt so acutely in the Midwest.

Some indirect diseconomies of centralization and large scale are much better known but equally hard to quantify. The long lead times of large units imply, for example, increased exposure to interest and escalation during construction, to mistimed demand forecasts (hence idle capacity), to wage pressures by strongly unionized crafts well aware (as in the Trans-Alaska Pipeline project) of the high cost of delay, to changes in political climate, and to changes in regulatory requirements. The last of these problems may be reinforced by elaborate assessment procedures required because large units concentrate social and environmental stress. Likewise during operation, the relative technical complexity of large systems entails longer downtime, more difficult repairs, higher training and equipment costs for maintenance, and higher carrying charges on costly spare parts made in small production runs.

The very conditions that make the indirect diseconomies of large scale important also make them hard to quantify. Nonetheless, some utility managers are realizing that interest, escalation, delays owing to greater complexity, and the effects of even slight forecasting errors over long lead times can make a single 1000 MW plant more costly than, say, five 200 MW plants or twenty 50 MW plants, all with shorter lead times and a broader base of engineering experience.

It is an inherent engineering feature of soft technologies, moreover, that their lead times are qualitatively shorter than those of conventional big systems. Whether in development, demonstration, or deployment, a small and technically simple system such as a rooftop solar collector, requiring months and thousands (at most) of dollars, is quicker and has lower indirect costs than one such as a fast breeder reactor, requiring several stages of scaling up at a cost of billions of dollars (and perhaps a decade) per stage. This scaling up process takes so long, indeed, that it is generally short-circuited by freezing the design and often beginning the construction of each new stage before any significant operating experience is gained with the previous stage. Gaining the experience would lose momentum and keep costly design groups idle, so there is an all but irresistible temptation to plunge ahead too quickly, so amplifying mistakes and cost overruns. In contrast, even a relatively adventurous soft or semisoft technology, such as a large wind machine, requires less than two years to design and build from scratch. The complete process for large solar heating systems today can be less than two months. The industrial dynamics of this process are very different, the technical risks smaller, and the dollars risked far fewer.

5.4. COMPARATIVE DEPLOYMENT RATES

Chapter 1.4 suggested that the relative deployment rates of hard and soft technologies are crucial for policy. Remarkably few data are available to resolve this point empirically. Relative technical simplicity, though, is clearly important. Consider this argument:

> . . . [T]he exponential growth of energy conversion in any industrialized country (excepting such special cases as Norway) is the sum of a series of overlapping exponential curves representing new energy sources, each introduced as the preceding curve matures or begins to falter, and each in general of a *simpler technical character* than those preceding. Thus coal displaced wood, and was in turn displaced by oil and by gas (the latter, with its exceptional simplicity [at large scale], accounting for 2/3 of U.S. energy growth from 1945 to 1965). . . . Continued exponential growth of total energy conversion requires that *each successive source be capable of more rapid growth than that preceding it*—possible only if there is a very large existing energy inventory capable of being cheaply and *rapidly* extracted [or captured], processed, and distributed, as was historically true of Persian Gulf oil and U.S. natural gas.
>
> Governments have apparently acceded to quick depletion of cheap energy reserves on the tacit assumption that a new source will turn up in time to maintain ever faster growth. *The latest such innovation, however, is nowhere in sight*, and a little reflection about the character that such a new source must have reveals why. Nuclear power, with its complexity and long lead times, is far too slow.[24]

Increasing technical simplicity at large total scale—the sort of progression exemplified by wood, coal, oil, and gas—is necessary to maintain growth, but is no longer available from hard technologies at the margin. And even the large-scale simplicity of natural gas has not been able to alter a disquieting trend: historical studies of the rate of

[24] See Chapter One, note 17. (Emphasis in original.)

market penetration, as plotted for example by Marchetti,[25] show that successive U.S. domestic fuel systems have gained a market share more and more slowly. This suggests that increasing technical simplicity at large scale in the wood-coal-oil-gas transition was more than offset by the increasing difficulty of keeping pace with growth in U.S. total energy use. Though market share charts do not say so, growth at 5 percent/y is vastly harder in a 1000 GW(e) grid than in a 10 GW(e) grid: it is the difference between adding 50 and 0.5 GW(e)/y, and that difference is logistically crucial.

Big energy systems based on energy capital have suddenly stopped getting simpler and have on the contrary become much more complex. Can the simplicity of smaller, softer systems offer a substitute? It is hard to give a *firm* theoretical answer, because soft technologies shift the burden from engineering logistics to social logistics, about which we know much less. But we do have impressive examples of significant innovations involving large numbers of small, relatively simple devices. For example, 54 percent of the houses in Vermont can now heat with wood stoves (not just fireplaces), and 6 percent have wood-fired central heating. Remarkably, 38–43 percent of all houses in Vermont, entirely through private initiatives, were backfitted with wood stoves during 1974–1976.[26] Another example: again entirely through private initiatives, the fraction of U.S. households attempting to grow some of their own food rose from 20 to 50 percent in 1971–1976, according to USDA statistics cited by Steinhart.[27] Significantly, both these innovations were highly decentralized, used local skills to install, and had some economic incentive. All but the last of these is also true of three phenomenal growth sectors of recent years: snowmobiles, pocket calculators, and citizens' band radio. The last, in early 1977, had reportedly reached over 13 million Americans, and the figure is rising by more than half a million per month.

It is hard to think of a really apposite example from the energy sector in an industrial country, but examples do exist in other countries. Many consistent reports[28] suggest that by late 1975 the People's Republic of China had deployed over a million (mainly village scale) biogas plants, of which 600,000 were reportedly installed in about a year and a half. Apparently more than 200,000 were installed in one

[25] See Marchetti, *supra* note 2, pp. 209–214.

[26] Gov. Thomas Salmon, personal communication, December 1976.

[27] J. Steinhart (University of Wisconsin, Madison), personal communication, January 1977, citing mid-1976 statistics.

[28] For example, by Denis Hayes (Washington, D.C.), Peter Hayes (Melbourne, Australia), and Dr. Ernst Winter (Katzeldorpf, Austria), based variously on reading of original literature, personal travel, and widespread interviews, all yielding broadly consistent results. See also the data of V. Smil, *Bull. Atom. Scient.* 33:2, 25–31 (February 1977).

year in Szechuan alone. Clearly the technical problems of such rapid deployment are not insurmountable even for a country without Western technical infrastructure. But the social structure also is clearly suitable, whereas in India the institutional problems of the villages have kept the deployment rate of biogas plants approximately one order of magnitude lower, with perhaps 45,000 units now working and a target of only 100,000 in 1980. These problems[29] — for example, the tendency of local elites to use energy systems (or food or land or water systems) to reinforce their own power, and the disparity between distribution of private capital and of energy needs—have some striking parallels in industrial countries such as the United States. This is not to say that we could not do anything quickly unless we had a Chinese political structure: even the recalcitrant institutional problems of the Indian village can be tackled by devising ways to give sufficient incentives to all the conflicting parties.[30] But it does mean that our logistical focus must shift from the technical to the social sphere, and we must be prepared to forget our classical theories of deployment rates (all developed in big, complex systems) and start from scratch with different rules.

Some of the most advanced industrial countries, notably Japan, have shown themselves capable of extraordinarily quick industrial changes on a vast scale. Western societies completely altered their industrial structure in six to nine months at the start of World War II. One of the most conservative U.S. sectors—housing—has recently shifted with startling speed toward mobile homes and condominia. But, as noted above, some industrial countries have also innovated rapidly on a small scale by not slowing private initiative with central bureaucracy. That may be the lesson of our limited experience with extensive small-scale innovation.

The steep downward trend of projections of future nuclear capacity in recent years—minus 15 to 40 percent per year in the U.S.

[29] For a classic account of village scale biogas plants in India, see C.R. Prasad et al., *Econ. and Pol. Weekly* (Bombay), 9:1347–64 (special issue, August 1974), and subsequent Errata; and A.K.N. Reddy, "The Technological Roots of India's Poverty" (Bangalore: Institute of Science, 1976). (The costs cited are three to seven times those of Chung Po's plastic bag digesters.) Institutional problems are considered in one way by Makhijani (see Chapter One, note 9, *supra*) and in another, less radical and perhaps more easily practiced, by J.K. & K.S. Parikh (IIASA), "Potential of Bio-Gas Plants and How to Realize It," in H.G. Schlegel & J. Barnea, eds., *Microbial Energy Conversion*, UNITAR seminar, Göttingen, 4–8 October 1976, pp. 555–91 (Göttingen: Erich Goltze KG, 1976). The Parikhs' approach is to distribute the economic benefits of the plant in such a way that all actors benefit—an approach analogous to that of the capital transfer scheme proposed for industrial countries in Chapter 2.7 as a way to co-opt utilities into the soft path. The stakes are high: a full-scale biogas program could by 2000 supply almost 90 percent of India's rural household needs (now about 45 percent of India's total energy use).

[30] *Ibid.*

since 1973—is due largely to prior neglect of logistical constraints in the social and economic sphere. The nuclear deployment rates still projected by many governments are generally faster than those projected in Chapter Two for much simpler, shorter lead time, politically benign soft technologies. Though it cannot be proved, it does seem plausible that one should be able to deploy simple technologies more quickly than complex ones; but the empirical evidence on this point is only suggestive, not conclusive. Conversely, though, people who say it is impossible to deploy soft technologies at the indicative rates given in Figure 2–2 must explain how the nuclear technologies on which their own projections rely can grow even faster. Nor are those solar deployment rates the highest in the professional literature,[31] even though every other published analysis assumes market penetration rates calculated on the basis of traditional price competition with present cheap fuels rather than with other long-run alternatives to them. A new comparison[32] shows that average deployment rates shown for soft technologies in Chapter Two are substantially lower than those projected by MITRE, Wolf, Morrow, and CONAES (though the comparison is not simple, owing to different definitions of technologies—soft versus all solar—and of how to compare the outputs of different sources). The soft technology assumptions in Chapter Two also bear comparison with the conclusions of the standard (though now somewhat dated) studies by NSF-NASA, USAEC, and FEA.[33]

5.5. SCALE IN SOCIAL CONTEXT

Not all qualitative scale issues are as subtly woven into the social fabric as are the crucially important effects considered in Chapters

[31] See the comparative graphs compiled by J. Weingart & N. Nakicenovic, "Market Penetration Dynamics and the Large Scale Use of Solar Energy," technical report in preparation (Laxenburg, Austria: IIASA, Spring 1977). For an up-to-date and highly revealing time-series graph of the steeply declining official U.S. nuclear projections, see C.F. Zimmerman and R.O. Pohl, "The Potential Contribution of Nuclear Energy to U.S. Energy Requirements" (Ithaca, New York: Cornell University, Department of Agricultural Economics, January 1977).
[32] Weingart & Nakicenovic, *ibid.*
[33] NSF/NASA Solar Energy Panel, "An Assessment of Solar Energy as a National Energy Resource," December 1972, reprinted in Part 1C, "Energy R&D and Small Business," Select Committee on Small Business, U.S. Senate (Washington, D.C.: USGPO, 1975); A.J. Eggers et al. (National Science Foundation), "Solar and Other Energy Sources," Subpanel IX Report to Chairman, USAEC, 27 October 1973 (Washington, D.C.: USAEC/USNRC Public Document Room, May 1974), excerpted in Small Business Committee, op.cit.; USFEA, "Project Independence Blueprint: Final Task Force Report: Solar Energy," November 1974 (Springfield, Virginia: NTIS), reprinted in "Loan Guarantees for Solar Energy Demonstrations," Committee of Science and Technology, USHR, no. 41, 7 October 1975 (Washington, D.C.: USGPO, 1976).

Two, Nine, and Ten. Some are very straightforward: for example, military vulnerability. The statement in Chapter Two that soft tech nologies are not militarily useful referred to offensive use; defensively they are extremely useful. The past few decades' military experience in Europe and Indo-China has taught us that central energy systems reliant on a few large facilities are far more vulnerable, and harder to restore when damaged, than dispersed systems. This concept has led to design criteria used today in the People's Republic of China and in Israel—which profited from the contrary approach of Syria when bombing the latter's oil facilities during the Yom Kippur War.

The adaptability of soft technologies to stress of all kinds can be viewed through a biological metaphor. Small energy systems adapted to particular niches can mimic the strategy of ecosystem development, adapting and hybridizing in constant coevolution with a broad front of technical and social change. Large systems tend to evolve more linearly, like single specialized species (dinosaurs?) with less genotypic diversity and greater phenotypic fragility. The adaptability of large systems is also constrained by the costly, specialized infra-structure they accrete. Such investments strongly influence future lines of development. For example, natural gas pipelines—by some reckonings of the third largest U.S. industry—provide, when not yet amortized, a strong incentive to fill them with synthetic pipeline-quality gas even at a price an order of magnitude higher. When the pipelines have been amortized, the incentive is still there because they can be considered a free component of a proposed synthetic fuel system. Likewise, building a grid dependent on GW blocks of electricity discourages a future shift to smaller scale or reduced electrification.

Small systems, in contrast, tend to depend on infrastructure not for distribution but for end use, thus increasing the user's ability to adapt. Resistive heaters and electric heat pumps are not very adaptable: an all-electric house is hard to heat except with electricity from some source, presumably a rather large source. But a domestic or district heating system based on circulating hot water can use virtually any heat source at any scale without significant changes in the domestic plumbing. It can adapt to as wide a range as solar collectors, solar/ heat pump hybrids, and combined heat and power stations. Thus, if a transitional technology such as coal-fired fluidized bed gas turbines with district heating (see p. 48 *supra*) is deployed first in those urban areas where solar backfits will be slowest and least convenient, with district heating perhaps clustered in subgrids with large holding tanks of neighborhood scale, then that interim heat distribution system (coupled to existing domestic plumbing) can be adapted later to whatever soft source of heat becomes available in each neighborhood.

Infrastructure at or near the point of end use can be designed for this sort of piggybacking, whereas large-scale distribution infrastructure enables one only to choose one or another kind of enormous power station, gas plant, etc.

Of course, such piggybacking saves long distance transmission costs without saving the local distribution costs, which are often larger (Chapters Six and Seven). But that does not mean that, for example, solar district heating systems are too costly to be worthwhile. The distribution costs will be already amortized, or mostly so, during the transitional era of fossil-fueled district heating (which can often recycle old municipal power stations through fluidized bed backfits). In the countryside, where marginal distribution costs might be very high, neighborhood or town systems would not be required because buildings are sufficiently spread out to collect their own solar heat without excessive shadow problems.

The object of matching soft technologies to end-use scale is to eliminate, as far as possible, the costs—including social costs—of distributing energy. This does not mean a shift entirely to small, decentralized systems, but rather a matching of the scale spectrum of supply to that of end-use. A shift of scale in a more decentralized direction will certainly be necessary, but not to an extreme. A few specialized industrial facilities will always require large-scale supply, which could be the large hydroelectric dams already in place in many countries.[34] As noted in Chapter 5.1, most end-use scale is much smaller than that; but there are often good technical and economic arguments for neighborhood scale. ("Neighborhood" has an operational definition that is social, not geographic: it is an area where people know each other, at least slightly, and treat each other as neighbors, so avoiding the high social costs of depending on people one does not know and get along with.) A neighborhood might include, say, of order 10^1-10^3 people. Neighborhood solar heating systems for single or cluster housing can clearly offer substantial economies over single house systems through freer collector siting and configuration, reduced craftwork,[35] reduced surface-to-volume

[34] Of course not all countries have such dams, and therefore not all countries could support, say, giant aluminum smelters; but not all countries can produce everything they want today. To say that West Germany, though she cannot grow rubber trees and coffee beans today, must nevertheless be able to smelt large amounts of alumina into the indefinite future is to advocate a rigidly ideological autarky beyond the wildest dreams of rural agrarian utopians. (On the other hand, some Alcoa executives interviewed several years ago believed it was quite plausible that in a few decades alumina would be smelted in large solar installations in the tropics where the bauxite and sunlight are—a unique application in which central station solar-thermal-electric stations might make sense.)

[35] This is important not so much because of cost as because the building industry is "dynamically conservative" in resisting innovation: R. Schoen et al., *New Energy Technologies for Buildings* (Cambridge, Massachusetts: Ballinger for The Energy Policy Project, 1975).

ratio in storage tanks, more favorable ratio of variable to fixed costs, and perhaps even a bit of user diversity (different people not all demanding heat or hot water is exactly the same pattern at the same time). Inspection of cities like New York suggests that virtually every area the size of a notional neighborhood is likely to contain at least one building that can accommodate enormously more collector area than it needs, sharing heat with its more shadowed neighbors. And in many cities the shadow problem is not nearly as important as one might think: the residual heat load of carefully insulated buildings can be made arbitrarily low, generally at a cost small compared to that of marginal supply. It is encouraging that some enthusiastic do-it-yourselfers have successfully retrofitted solar water heaters even into tall Manhattan brownstones.

The same spectrum of scale would apply to transitional as to the ultimate soft energy systems. It is even too early to rule out entirely the possibility of fairly large-scale coal conversion to make industrial gas for a transitional ring main in the Midwest, until fluidized bed backfits (especially for industrial packaged boilers, with or without cogeneration) can take up the slack. For the time being, most such boiler backfits will also have to match existing industrial scale whether it is optimal in the long run or not. But all such transitional systems, unlike those now planned, would be designed at appropriate scale so that they and their infrastructure can be shared: cities now building subways, for example, should leave room for district heating pipes in the tunnels.

Large-scale energy systems already in place (not at the margin) can be useful transitional technologies on a national or regional scale. For example, my studies of a soft energy path for Denmark[36] came to conclusions similar to Sørensen's,[37] though I assumed more organic conversion, less electricity, and no hydrogen—that is, a more careful matching of energy quality to end use needs. But the fifty-to-seventy-five year transition, even though no more awkward than under present policy (and probably much less so), could be made even easier within a Scandinavian hegemony that shared surplus Norwegian hydroelectricity in return for food, steel, and other exports from Denmark and Sweden. Likewise, liquid fuels from the French biomass base might be usefully shared with the Low Countries in return for local wind-based electricity, or Scottish hydroelectricity (and perhaps wavepower) swapped for English manufactured goods. Current studies within the EEC[38] may reveal scope for other such regional

[36] See Chapter One, note 18, Introduction.
[37] See Chapter One, note 25 (original *Science* article, with graphs slightly revised in Danielsen et al.).
[38] Supervised by Professor Niels Meyer, Physics Lab. III, Technical University of Denmark (Lyngby).

policies—though they are not an argument for building more big systems than we have now.

The concept of matching scale to end use might be illustrated by the familiar analogy of transport. The amount of transport required to gain desired access, like the amount of energy required to perform desired tasks, can be reduced by technical and structural changes, but nonetheless the diversity of settlement patterns and lifestyles in a pluralistic society will yield a spectrum of transport densities needed. Mass transit, often electrified, is most appropriate for high density intra- and intercity commuting; variable route public transport (such as dialable buses), mixed with both more and less personal forms of transport, is most appropriate for medium density suburbs; private vehicles (efficient cars, scooters, bicycles, etc.) and a sprinkling of other modes are most appropriate in the countryside. The essence of rational transport planning is to mesh mode and need in this way, rather than trying to bring a dense subway network to the countryside or private cars to the inner city. Just so, a rational energy system uses small solar collectors for single houses and large existing hydroelectric dams for smelters—not the other way around, which would be a symmetrically nonsensical mismatch.[39]

It is this matching of the spectrum of energy and transport density supplied to that required that is the essence of the soft approach—not the predetermined predominance of a particular scale, small or large. That is why conservationists are consistent in desiring both generally smaller energy systems and more urban mass transit, and why their adversaries, with equal consistency, tend to desire both gigantic energy systems and the universal primacy of private cars. It happens that most of our travel is in and between cities, and that the balance should therefore shift toward appropriate urban transport modes and away from private cars. Likewise it happens that the current mismatch of scale between energy supply systems and end-use needs entails a substantial shift toward smaller scale in most cases. But these shifts are a consequence of a general soft path principle of matching and appropriateness, not the other way around. Each energy or transport system should do what it does best, not try to do something else or be a panacea.[40] In this sense it is the hard path, with its

---

[39] This is precisely the mismatch contemplated by those who propose to heat houses with fast breeder reactors, yet ask how wind machines can operate smelters—especially smelters built to the unnecessarily large scale made possible by subsidized dams.

[40] In a sense this discussion restates the traditional argument over how widely to cast the net of economic costs and benefits. A traditional economist might consider only a narrow set of pecuniary costs and benefits and so conclude that GW(e) scale, or its equivalent in nonelectric energy technologies, is

bias toward homogeneity and large scale, that is ideologically rigid, and the soft path, with its emphasis on appropriateness to the task at hand, that reflects a flexible, pluralistic social fabric[41] .

appropriate for virtually every application. This chapter merely suggests that internalizing a wider set of costs and benefits, especially social ones, would yield a different result for most energy systems in most circumstances. Implicit in this argument is the idea that intangible social costs of the kind considered in Chapters Two, Nine, and Ten are not separable from traditional internal costs, but are inextricably linked and must be considered together.

[41] Scale issues are universal in modern technologies. They were noticed early in the chemical industry: see, e.g., U. Columbo and G. Lanzavecchia, "Criteria di scelta della potenzialità degli impianti chemichi," *Ing. Chim. Ital. 9*:2, 22–30 (February 1973). Professional fee structures and education, oligopolistic forces, and intellectual sloth have likewise led to a trend toward giant sewage treatment plants. These tend to be technically inflexible and inelegant, vulnerable to strikes and large-scale technical failures, not amenable to backup, and dependent on a costly collection system (distribution backwards). Often the institutions responsible for the gigantism do it rationally because they do not have to pay for its diseconomies, or do not perceive them because of limited terms of reference. The same is true in transport, industry, cities (I. Hoch, *Science 193*:856–63 [1976]) and administration (see Chapter 10, note 9). A fundamental reassessment of scale issues throughout our lives is long overdue, and is a never-ending task, for as societies change, appropriate scale changes too.

Chapter 6

Capital Costs of Hard Technologies

6.1. CENTRAL ELECTRIC SYSTEMS

This chapter will compute the approximate marginal capital costs of complete hard technology energy systems, subject to the caveats and in the methodological spirit of Chapter 3.2. The data base for this chapter is that of the Bechtel Energy Supply Planning Model,[1] expressed in third quarter 1974 U.S. dollars for equipment ordered in early 1974. The data base, like any, could be improved, but is probably the most detailed, authoritative, and up to date available (it is being continuously updated[2]), and is widely used in federal and private energy studies. The data base will be used to compute capital costs first for nuclear-electric and then for conventional centralized coal-electric systems in the U.S. Unfortunately, such computations do not yet appear to be possible in other countries, owing to lack of adequate published data or lack of a fair approximation to a genuine market in equipment or both.

The Bechtel estimate[3] of direct construction cost—architect-engineer's pass-through—for a 1.1-GW(e) light water reactor power

[1] M. Carasso et al., *The Energy Supply Planning Model*, Bechtel Corp. report to NSF, PB-245 382 and PB-245 383 (Springfield, Virginia: NTIS, August 1975).

[2] M. Carasso & J.M. Gallagher, personal communications (1976). Dr. Carasso, now with the California Energy Resources Conservation and Development Commission, has kindly reviewed my use of his group's data (*supra* note 1) to ensure I was using it correctly. Dr. Gallagher has generously shared the results of Bechtel's update of the model's data base, an effort he now directs.

[3] *Supra* note 1.

station is \$418/kW(e) installed (net of on-site electricity requirements).[4] To this cost must be added[5] 40 percent for indirect or owner's costs—interest during construction, design, and administration—for a total of \$585/kW(e) installed. This is comparable to the \$520/kW(e) (also 1974 dollars) assumed by the Brookhaven data base,[6] which relies on the out of date (July 1973) and poorly documented A.D. Little study for Northeast Utilities.

Associated with a marginal GW(e) of generating capacity in a typical U.S. program is about 140 miles of transmission circuit whose nominal characteristics[7] are shown in Table 6–1. The right-hand column, calculated from the other data given, implies that the total cost of transmission equipment (direct plus owner's costs, 1974 dollars) associated with a marginal kW(e) of generation capacity is about \$69. (The corresponding 1973 and 1985 average costs are respectively about \$87/kW and \$80/kW, internally consistent within rounding errors.) The Brookhaven estimate[8] of \$75/kW (1974 dollars) is fairly close. Note, however, that the current dollar escalation rate averaged about 11 percent/y during 1966–1972, when GNP inflation was relatively slow and steady,[9] and that regional cost variations are about 3.5-fold.[10] The Bechtel data,[11] based on extensive field experience and checked with utility and FPC data, will be assumed authoritative.

The Bechtel estimate[12] of capital cost for distribution equipment assumes a nominal facility with 160 MVA input. It feeds 29 MW, 21 MW, and 70.2 MW to industrial 34.5 kV, commercial 13.2 kV, and domestic 220 V customers respectively, at an average power factor such that MVA x 0.823 = MW. Direct construction cost[13] in 1974 dollars is M\$34.5 with all lines aerial, M\$55 with all lines underground, and M\$41, or k\$256/MVA, for a nominal split (about 80 percent underground) considered realistic at the margin.[14] Indirect costs[15] are 35 percent. Total capital cost of marginal distribu-

---

[4] Many current data, e.g., the Rancho Seco projections (see Chapter Three, note 20), strongly suggest that this figure is already 10–30 percent too low.

[5] *Supra* note 1.

[6] M. Beller, ed., *Sourcebook for Energy Assessment*, BNL-50483 (Springfield, Virginia: NTIS, December 1975).

[7] *Supra* notes 1 and 2.

[8] *Supra* note 6.

[9] See Chapter Five, note 5.

[10] *Ibid*.

[11] *Supra* notes 1 and 2.

[12] *Ibid*.

[13] *Supra* note 2.

[14] *Ibid*.

[15] *Supra* note 1.

Table 6-1. Nominal Characteristics of U.S. Marginal
Electrical Transmission Facilities[a]

| Voltage | Capacity (MVA/facility) | | Length (GW^b-mi/ | Cost (M\$^{c}/ | Approximate Modal Split[d] | | |
	in	out	facility)	facility)	1973	1985	marginal[e]
230	600	530	125	95	0.47	0.32	0.185
345	1200	1060	300	120	0.23	0.27	0.300
500	2600	2450	600	188	0.23	0.29	0.335
765	4000	3800	1250	228	0.05	0.09	0.120
400 DC	1400	1370	1200	317	0.02	0.04	0.055

[a]See notes 1 and 2, *supra*.
[b]GW transmitted; to express in terms of GW generating capacity, divide by 0.791. This factor expresses "supply diversity," i.e., the ability of several sta tions to share a [statistically] reduced amount of transmission capacity because they are unlikely all to be producing full power at the same time.
[c]Direct construction cost, third quarter 1974 dollars; for owner's costs add 20 percent.[f]
[d]Fraction of total energy transmitted per mode (not fraction of facilities per mode). Columns may not total 1.00 due to rounding.
[e]Computed by the author for the 1973-1985 period, assuming[f] total generating capacity of 415 GW in 1973 and 899 GW in 1985.
[f]See note 1, *supra*.

tion is thus about \$420/kW (1974 dollars).[16] Since the *Electrical World*-Brookhaven estimate[17] is only \$145/kW, the Bechtel es-timate[18] has been queried with special care. Reanalysis by Bechtel has confirmed,[19] as explained in a Bechtel memo to Brookhaven, that the Bechtel \$420/kW is realistic. The discrepancy arises, in Bechtel's view, from utilities' less stringent undergrounding assump-tions and special internal accounting procedures that reduce apparent overheads and cost of money.

The Bechtel estimates,[20] also used by Brookhaven,[21] for marginal capital costs of nuclear fuel cycle facilities yield a total cost of about \$61/kW(e) installed (1974 dollars). This figure appears to be too low by a substantial factor[22] (probably at least two) for 1976 ordering, but is assumed here anyhow.

[16] The original error band estimate (see note 1, *supra*) is minus 10 percent, plus 20 percent, with an exponential scaling factor 0.7.
[17] *Supra* note 6.
[18] *Supra* note 2.
[19] *Ibid.*
[20] *Supra* note 1.
[21] *Supra* note 6.
[22] J. Harding (then on staff of CERCDC [see note 2, *supra*]), personal com-munication, 1976; Harding is among the best-informed authorities on fuel cycle economics.

A fairly up to date estimate[23] of the approximate capital cost of the initial reactor core—a front end cost even though it can be credited later against fuel cycle costs—is given in Table 6-2 (in 1976 dollars, not 1974 dollars, for 1976 ordering). It shows that the initial core costs about $100/kW(e) installed (1976 dollars).

The 1974-dollar costs must now be converted to 1976 dollars. Using a standard index for general construction, such as Handy-Whitman, is a cruder method than available data can justify. Such a method is probably appropriate for transmission and distribution facilities, and, in the absence of more detailed studies, for fuel cycle facilities; the index used here, as by Brookhaven,[24] is the Marshall and Stevens Equipment Cost Index, which for 1974-1976 is 1.25. The total cost of these three components in 1976 dollars is thus 1.25 x ($69 + $61 + $420) = $688/kW(e) installed.

A detailed and authoritative cost escalation study is available for U.S. light water reactors[25]; it is far more persuasive than the virtually useless and undocumented escalation rates given in official analyses.[26] The Bupp and Treitel simple regressions[27] yield capital cost escalation rates of

14 percent/y in current dollars or 7½ percent/y in constant dollars for LWRs,

13 percent/y in current dollars or 6½ percent/y in constant dollars for coal-electric plants,

but the fit, with $r^2 = 0.16$, is inferior to a multiple regression ($r^2 = 0.71$) that includes dummy variables for multiunit and Northeast region siting. The multiple regression yields escalation rates of

26 percent/y in current dollars or 20 percent/y in constant dollars for LWRs,

13 percent/y in current dollars or 7 percent/y in constant dollars for coal-electric plants.

Applying the simple and multiple regression results to the 1974-dollar power station costs cited earlier—but *only* over the period

[23] *Ibid.*

[24] *Supra* note 6.

[25] I.C. Bupp & R. Treitel, "The Economics of Nuclear Power: De Omnibus Dubitandum," 1976 typescript from Professor Bupp at Harvard Business School. These January 1976 regressions treat a very detailed data base (see Chapter Three, note 19) for thirty-five U.S. LWRs and 46 U.S. coal plants complete or nearly complete in summer 1974. Professor Bupp's more recent studies suggest that similar escalation rates still prevail today.

[26] E.g., USAEC, WASH-1345 (1974).

[27] *Supra* note 25.

Table 6-2. Approximate Capital Cost of the Initial Core of a 1 GW(e) PWR[a,e]

Item	Unit Cost	Cost Per Annual Core	Cost Per Initial Core
400,023 lb U_3O_8	$42/lb	M$16.8 $\times$ 3 =	M$50.4
153,838 kg U to convert to UF_6	$3.50/kg	0.5 $\times$ 3 =	1.6
135,068 kg U to enrich to 27,001 kg 2.63%^{235}U with 74,685 SWU	$67.25/SWU	5.0 $\times$ 3 =	15.1
28,994[b] kg U to fabricate	$90/kg	2.6 $\times$ 3 =	7.8
TOTAL DIRECT COST			M$74.9/GW(e)
interest at 12 percent/y^c			24.4
TOTAL COST^d			M$99.3/GW(e)

[a]1976 dollars and ordering; initial design enrichment 2.63 percent; tails assay 0.3 percent; process losses included; thermal efficiency 0.325; capacity factor 0.55; burnup approximately 22 GW(t)-d/TU^{35}.

[b]Larger owing to complex correction for internal process recycle.

[c]Assuming lead times of 3 y for U_3O_8 production and conversion, 1.5 y for enrichment, and 0.6 y for fabrication.

[d]Excludes insurance, property taxes, and other minor overheads. The fuel cycle facilities whose sunk costs are being amortized in the unit costs shown in the second column are historical facilities, whereas those counted separately in the text are marginal facilities to be built to supply new reload fuel; thus facilities are not being double counted.

[e]See note 22, *supra*.

1974–1976, *not* thereafter—yields a 1976–dollar power station capital cost of $761/kW(e) and $929/kW(e) installed, respectively. Both these values will be used in later calculations.

The total marginal capital cost in 1976 dollars for an installed kW(e) of the nuclear system is thus $688 + $100 [initial core] + ($761 or $929) = $1548 or $1717, depending on which 1974–1976 index is used for the power station. This total cost must now be converted to $/kW(e) sent out by dividing by an appropriate capacity factor. The empirical capacity factor of the U.S. LWR inventory up to mid-1976 averages about 0.58.[28] Some analysts argue that this will improve on a "learning curve"; others question whether this theory is applicable to the circumstances peculiar to nuclear engineering with possibly declining quality control,[29] and

[28] See Chapter Five, note 21.
[29] The doubling time of reactor population, before the recent hiatus in ordering, was far shorter than the doubling time of the most competent and dedicated quality control engineers, especially those left over from the early days of the naval reactor program.

note that the infirmities of LWRs appear from outage analysis to be proceeding with scarcely a pause from the pediatric to the geriatric. Komanoff's exhaustive multiple regressions, doing the best that is possible with present data to separate the effects of age and of unit size, suggest that increasing unit size will make capacity factors decline. He estimates a ten year levelized average capacity factor of about 0.50 for a new 1.1-GW(e) PWR,[30] a value that is arguably conservative (too high). In contrast, the lifetime average capacity factor (not levelized) assumed by the USAEC is 0.57.[31] Experience abroad is ambiguous (West Germany versus Japan). In the absence of better data—which should emerge in the next few years—a round-numbered capacity factor of 0.55 is assumed here. Moreover, since definitive data are not available on whether the corresponding capacity factor for marginal transmission and distribution equipment should be slightly higher or lower than for the power station,[32] the same average figure will be used for the entire system to convert from kW(e) installed to kW(e) sent out. The total cost (1976 dollars) per kW(e) sent out is thus $2815 or $3121.

Finally, to convert from kW(e) sent out to kW(e) delivered, we must take account of transmission and distribution losses, assumed here to total 10.7 percent. This figure—much lower than the 8.4 percent and 8.7 percent respectively (16.4 percent cumulatively) assumed by Bechtel[33]—is accommodated by an equivalent surcharge of 12 percent on capital cost. The result is $3152 or $3496 per kW(e) delivered, depending on the index used to convert power station cost from 1974 dollars to 1976 dollars. We have thus calculated, in 1976 dollars, the total marginal capital cost of a nuclear-electric system.

But this result is highly conservative (too low) because it omits some significant costs that ought properly to be within the system boundary. These omitted costs include marginal capital investment in:

[30] *Supra* note 28.

[31] USAEC, WASH-1139(74), 1974, p. 23.

[32] This would matter for a particular facility, but should not matter for the notional marginal facility considered here, since it is a national average. Local user or supply diversity should thus "wash," as the econometricians say. Supply diversity should certainly not be counted here for the transmission network, as that would be double counting (see Table 6-1, note b). A completely different approach to the type of calculation considered here would be to apply appropriate loss corrections to the combination of a marginal kW of generating capacity plus its assumed fractional kW of reserve margin. Still other methodologies could be used to obtain the same result; the approach used here is simplest to understand.

[33] *Supra* notes 1 and 2.

land;
reserve margin and spinning reserve[34];
taxpayer-supported regulation and security services;
federal research and development (appropriately amortized);
future services (waste management and decommissioning); and
end-use devices.

No allowance is made for cost escalation after 1976 at any rate in excess of the GNP inflation rate; this seems, in the light of Bupp and Treitel's results above, a highly conservative assumption. (Escalation at 20 percent/y in constant dollars for only six years would treble the cost.) No allowance is made for the energy that must be debited against station output to operate the nuclear fuel cycle—a debit of 6-8½ percent for most LWRs.[35] A realistic calculation including all these terms except end-use devices, and still excluding all externalities and all dynamic net energy considerations,[36] would probably yield a nuclear capital cost nearer $300,000/(bbl · day) delivered enthalpy than the $211,500 or $234,600/(bbl · day) derived above. The numbers at this point obviously become fuzzy because proper data are not available and, in some cases, may never become available. Nonetheless the estimate of $3150-3500/kW(e)

---

[34] A.B. Lovins, "Things that go pump in the night," *New Scient.* 58:564 66 (31 May 1973). True to form, the M£100+ Dinorwic scheme was both overrun and obsolete before it was even off the drawing board, but is being built anyway. The economic details are in the Commons Private Bill Committee hearings, especially in the author's evidence, and in more recent trade press accounts of the overrun (now reported to exceed a factor of two).

[35] ERDA-76-1, Appendix B (1976): the static analysis only, as the dynamic analysis appears, from the limited information given, to be wrongly done. The static analysis is a factor of 1.7 less favorable to PWRs than Price's (see Chapter One, note 18): normalizing to Price's accounting conventions, the ERDA P_u/P_i is 1.97. Price's epilogue gives a range of other estimates. It is also important to note that all analyses assume that the design basis fuel burnup is actually achieved. A large fuel sample analyzed by Harding (see note 22, *supra*) in responding to interrogatories in the LILCO Jamesport hearings in mid-1976 suggests that only about half to two-thirds of the design basis burnup is being achieved by mature LWR cores. If this is correct—and informal checks around the industry confirm it—then the reactors are consuming yellowcake and separative work far faster than standard calculations of fuel cycle economics (and uranium supplies) assume. (ERDA is surveying the burnup data, but so far no results are available.) This revelation does not improve the outlook for uranium supply. Nor can fast breeder reactors help much, for reasons explained by L. Grainger, *Energy Policy* (U.K.) 4:322-29 (December 1976), and D. Merrick, *Nature* 264:596-98 (16 December 1976); *cf.* C.E. Iliffe (UKAEA), "Economic and Resource Aspects of Fast Reactors," 15 January 1977. An elegant dynamic model of this problem appears in Roger Brown's January 1977 M.S. thesis (System Dynamics Group, Dartmouth College, Hanover, New Hampshire).

[36] See Chapter One, note 18.

delivered worked out above seems unrealistically low: $5000 might not be too high.

Of course all these data can be refined and improved—though it may not be worth working hard on second and third significant figures in view of the major unknowns in future escalation rates. The main defect in the approach illustrated seems rather to be that it does not take account of generating mix (for different parts of the load duration curve) or of user diversity.

On the former point, clearly gas turbines, combined cycle plants, etc., are less capital-intensive than baseload stations—though more costly in fuel, which would tend to compensate if, as suggested in note 26 of Chapter Three, the stream of future fuel costs were present valued and combined with capital costs as a basis for comparison with soft energy systems. But the effect of plant mix on total investment is diluted anyhow by the high cost of transmission and distribution. The latter term especially dominates and is not greatly reduced by siting plants such as gas turbines near load centers. Moreover, detailed data on the capital costs of complete electrical systems at the margin, including a realistic generating mix, appear not to be available. The only recent number of this type is an estimate[37] of $2000 capital cost per average kW of capacity to deliver to Con Ed customers in New York (where land and underground cable costs are high). This number appears to refer to capital cost per kW installed, not sent out or delivered, and should therefore presumably be roughly doubled to reflect capacity factor and distribution losses. In the absence of better data, it is somewhat reassuring to reflect that most official long-term projections of generation mix assume that nuclear capacity will extend far beyond the present baseload level and that much of the current intermediate load and gas turbine territory will be occupied by very capital-intensive pumped storage schemes.[38] If this were done, the objection considered in this paragraph would disappear, since pumped storage schemes are considered by their proponents to compete with gas turbines only because the zero fuel costs and moderate losses of pumped storage systems more than compensate for their higher capital costs compared to gas turbines.

As for user diversity—the fact that not all users will demand electricity at once, so they can share capacity—it is a real issue on which better data are needed. It is not particularly relevant, though, to the specific (and, for marginal electrification, crucially important)

[37] J.J. O'Connor, "The $2000 Kilowatt," *Power* (April 1976, p. 9).
[38] *Supra* note 34.

application to be examined in Chapter 8.1, space heating, since European experience clearly shows that the winter peak heating load determines the system's simultaneous maximum demand with negligible user diversity. For industrial process heat (which, from resistance heaters, accounts for 60 percent of the marginal electricity use in ERDA's 1975 "Intensive Electrification" scenario,[39] or more electricity than the U.S. used for everything in 1975), one can again probably assume a good approximation to base loading, so the question does not arise. For intermittent applications such as home appliances, it clearly does arise and should be taken into account; but the terms to which a large correction factor would apply are only small terms in the total electricity budget.

The nuclear system costs calculated above deserve comparison with equivalent coal-electric system costs: not because this is the only other way to meet end-use needs, but because both are more capital-intensive than many other energy systems considered in Chapters Seven and Eight. The Bechtel data[40] for 800 MW(e) coal-electric plants, averaged over the various types of coal and of coal mines and delivery systems in a typical U.S. hard technology program, show an average capital cost of about $355/kW(e) installed (1974 dollars, direct plus owner's costs). About $106/kW(e) must be added for the fuel cycle and $120/kW(e) for scrubbers.[41] Assuming 1974-1976 inflation for the power station and scrubbers at about 13 percent/y, according to the multiple regression used earlier,[42] and assuming the same transmission and distribution costs and 1974-1976 indexes as assumed above for nuclear systems (but a liberal 20 percent/y for the coal fuel cycle during 1974-1976), yields a total capital cost of $1370/kW(e) installed (1976 dollars). A predicted capacity performance of 0.62[43],[44] and a 12 percent surcharge for transmission and distribution losses, as above, yield

---

[39] F. von Hippel & R.H. Williams, "Energy Waste and Nuclear Power Growth," *Bull. Atom. Scient.* 32:10, 18 ff. (December 1976). A similarly devastating critique of British plans for electrification has been published as ERG 013 by P.F. Chapman et al. (Energy Research Group, The Open University [Milton Keynes, Bucks., U.K.], 1976), and is summarized by D.A. Casper, *Energy Policy* (U.K.) 4:191-211 (September 1976).

[40] *Supra* note 1.

[41] *Ibid.*

[42] *Supra* note 25.

[43] *Supra* note 28.

[44] Capacity performance is a fair representation of what a station (in this case, a supercritical station with scrubbers) would achieve as an average capacity factor if it were baseloaded. If, as Komanoff shows even with highly conservative fuel cycle costs, coal-electric stations send out cheaper electricity than LWRs (see note 28, *supra*), then coal stations at the margin should be baseloaded and this computation of capacity performance is unnecessary.

a total cost of \$2476/kW(e) delivered (1976 dollars), equivalent to \$166,100/(bbl · day) delivered enthalpy. The omitted terms are analogous to, but presumably smaller than, those for nuclear power as summarized above.

Neither of these calculations is, strictly speaking, a whole system cost, not only because of their noted omissions, but also because neither calculates the cost of supplying a unit of end-use *function*, so taking account of the high quality of the form of energy produced. This point is considered in Chapter Eight. The end-use efficiency of electricity cannot be generalized, but must be considered separately for each end use. Any attempt to anticipate that process would make it impossible to derive a general data base (for the wide range of technologies considered in Chapters Six through Eight) that could then be applied to any end use.

6.2. DIRECT AND SYNTHETIC FUEL SYSTEMS

Direct fossil fuel systems are surveyed in detail by Bechtel,[45] with the results shown in Table 6-3. The Brookhaven model[46] uses the same data. The values marked with an asterisk in Table 6-3 assume an average weighted by the modal mix (both for sources and for means of transport) calculated by the Bechtel model[47] as least cost, assuming the January 1975 State of the Union Message energy program for 1976-1985. Since the delivery system costs are a function of the source, the detailed structure of the distribution system should really be calculated separately in each case. In general, the Arctic sources, with low wellhead investment, entail high transport investment (an Arctic gas pipeline ordered now could easily cost much more[48] than the associated wellhead investment). The exact data, insofar as anyone knows them, are in principle available from the Bechtel data base,[49] but would require very laborious calculations of modal splits. As a crude approximation, therefore, one can only combine weighted averages of sources in each category with weighted average costs of delivery. The result in 1976 dollars, order-

[45] *Supra* note 1.
[46] *Supra* note 6.
[47] *Supra* note 1.
[48] H.R. Linden (President, Institute of Gas Technology), information submitted for the record, Committee on Science & Technology, full Committee meeting on loan guarantee provisions of HR 3473, ERDA Appropriations Bill, FY 1976, USHR, 18 Sept. 1975; see also in *En. Systs. & Policy 1*:325-51 (1976). Interesting comparative data on escalation and on electric *vs.* synthetic conversion of coal are included in both versions.
[49] *Supra* note 1.

Table 6-3. Marginal Capital Costs of Some Complete U.S. Direct Fossil Fuel Systems[a,b] ($/(bbl · day))

System	Cost of Exploration, Development, and Production, at Wellhead, Minemouth, or Dock	Cost of System to Process and Deliver to Consumer
Domestic coal: weighted average	$ 1,437*	
Eastern underground	1,150	
Eastern surface	1,980	$1,380*
Western underground	968	
Western surface	1,150	
Crude oil import	110	
Alaskan oil	803	
Offshore lower 48 oil	11,594	3,697*
Oil shale	9,823	
Liquefied natural-gas import	1,034	
Alaskan gas	3,410	
Offshore lower 48 gas	11,594	4,698*
Onshore lower 48 gas	12,496	

[a]Direct plus owner's costs; 1974 dollars; early 1974 ordering; includes process losses and the costs and losses of delivery to consumer.
[b]See note 1, *supra*.
*See text.

ing in 1976, is approximately $2800 per delivered bbl/day for domestic coal, $7200 for the oil sources shown, and $13,100 for the gas sources shown. Both of the latter figures are significantly lowered by the inclusion of imports (imported oil is only about as capital-intensive as domestic coal, a few k$/(bbl · day)). It appears that domestic frontier fluids fall generally in the range k$10-25/(bbl · day) in 1976 dollars.

As for coal synthetics, the Bechtel data[50] relied on by the Brookhaven model[51] are in the range (1974 dollars and ordering) from k$14.9/bbl · day) for industrial gas to k$20.7 for pipeline quality gas and k$34.3 for methanol. (Brookhaven estimates[52] k$12.5 for hydrogen.) These estimates do not include, however, the fuel cycle, water supply, or delivery system (though the last of these might rely on pipe already laid). Broadly comparable estimates are given in the November 1975 Interagency Task Force report on synfuels. September 1974 data[53] suggest that the specific investment (1975

[50] *Ibid.*
[51] *Supra* note 6.
[52] *Ibid.*
[53] *Supra* note 48.

dollars) in a Lurgi plant is of order k$18/(bbl · day) for Western and k$20 for Eastern coals, with about a 20 percent saving possible from second generation (e.g., HYGAS) technology. The total cost for such an advanced plant, including the fuel cycle, water supply, and connections to existing pipelines, would be of order k$22/(bbl · day) "or more," roughly comparable to the cost of a tar sands system.[54] The extraordinary escalation of the cost of most synthetic fuels projects during 1974-1976, however, makes it seem likely that a whole system capital cost approaching k$40/(bbl · day) in 1976 dollars, ordering in 1976-1977, is probably realistic, and might even be low if such rapid escalation did not stop promptly. Squires[55] estimates at least k$40/(bbl · day) for delivered synthetic gas, assuming that the pipelines are free.

No reliable data are available for oil shale, as there are major uncertainties in process requirements and in the supply of water and the disposal of solid and liquid wastes. Barring unforeseen progress with *in situ* recovery, estimates comparable to those for coal-based synthetics seem plausible. Boucher[56] estimates a capital cost of order k$20/(bbl · day) for marginal surface tar sands in Alberta, though small increments of production might be possible at much lower marginal cost by removing bottlenecks in the present syncrude system.[57]

[54] *Ibid.*

[55] A.M. Squires, "Recapturing Control of Our Clean Fuel Supply," CEQ Hearings on the ERDA RD&D Plan, Austin, December 1976.

[56] R. Boucher, "Commentaires sur Certains Aspects d'une Societé de Conservation pour le Canada" (Montréal: Ministère des Richesses Naturelles, 16 August 1976).

[57] D. Bowie (Petro-Canada), personal communication, November 1976.

Chapter 7

Capital Costs of Transitional and Soft Technologies

7.1. TRANSITIONAL FOSSIL FUEL SYSTEMS

For some transitional fossil fuel technologies, such as the flash conversion processes[1] and supercritical gas extraction,[2] no

---

[1] See, e.g., A.M. Squires, *Science 184*:340-46 (1974), *189*:793-95 (1975), *191*:689-700 (1976); and "The City College Clean Fuels Institute" (Paper delivered at symposium on "Clean Fuels from Coal," Institute of Gas Technology, Chicago, 26 June 1975 (this paper also describes panel-bed filters); J. Yerushalmi et al., *Science 187*:646-48 (1975) and *I&EC Process Design & Devel. 15*:47-53 (January 1976) (Amer. Chem. Soc.). The flash processes—both pyrolysis and hydrogenation—do not get the coal hot enough for long enough to make messy, corrosive, and carcinogenic tars.

[2] D.F. Williams, *Applied Energy 1*:215-21 (1975) (Applied Science Publishers, U.K.); K.D. Bartle et al., *Fuel 54*:226-35 (October 1975) (IPC, U.K.). The process exposes lumps of coal, without special heating, to solvent vapor above its critical pressure, e.g., toluene at 100 bar. The vapor penetrates between the leaves of carbon and pulls out intact (usually aromatic) molecules by the interstitial hydrogen without cracking them. The hydrocarbons then fall out of the vapor when it is reexpanded in another container. A substantial fraction of the mass of the coal is thus extracted in a form (typically benzene, anthracene, etc.) whose spectrum is controllable by the choice of solvent, pressure, and exposure time. Virtually all the hydrogen is removed. Metals may be removed too (this is still uncertain). The residual char—carbon plus minerals—generally melts in the range 100-200°C and could be liquid fed as an underboiler fuel. Like the flash processes (see note 1, *supra*), supercritical gas extraction offers the intriguing possibility of small units sited in a coal-using factory; premium fuels extracted on the premises could then be used for, say, kiln firing and the char for cogeneration and process steam (see note 11, *infra*). This flexible scale, simplicity, and lack of tar formation seem very attractive compared to conventional G$ conversion plants, but the process is little known outside England.

firm cost estimates appear to be available. These technologies do, however, seem simpler and more reliable than conventional large-scale conversion (Chapter 6.2), and it would be surprising if they did not have correspondingly lower capital costs and escalation rates.

Firmer estimates can be given for the fluidized bed technologies on which Chapter Two proposes strong and immediate commercial emphasis.[3] Tender prices by several confident vendors for the 25 MW(t) multifueled (oil-gas-coal) fluidized bed boiler to be commissioned in late 1977[4] by Kymmene-Mustad (the latter of whom now operate a 2 MW version) for Enköpings Värmeverk[5] are still confidential. Fortunately, a detailed design study by Stal-Laval Turbin AB (Finspong, Sweden) for another type of fluidized bed system (Chapter 2.6) has been developed to the point of cost estimates on which a commercial tender could be based,[6] and these open literature values permit some useful comparisons.

[3] See the citations in Chapter Two, notes 26-28; A.M. Squires, "Applications of Fluidized Beds in Coal Technology," in J.P. Hartnett, ed., *Alternative Energy Sources* (Washington, D.C.: Hemisphere, 1977), perhaps the best tutorial available; the brochure "Fluidized Combustion," Combustion Systems Ltd. (66-74 Victoria St., London SW.1), which is amply illustrated for a nontechnical audience; H.B. Locke & H.G. Lunn, *The Chemical Engineer* (November 1975, pp. 667-70); and the general review by A.M. Squires, *Ambio 3*, 1:1-14 (1974). A fuller description of domestic fluidized bed furnaces (see Chapter Two, note 27) might be useful. They are very compact: a unit producing 8.8 kW(t) (30kBTU/h) is about $25 \times 25 \times 36$ cm. From the user's point of view it should be indistinguishable from an oil furnace (except cheaper). Granular coal is delivered by hose from a tank truck to a bunker, then gravity fed to a small packaged furnace with very low emissions (NO_x typically a few parts in 10^6). Ash is collected by overflow in a snap-on container akin to a vacuum cleaner bag, changed at each fuel delivery, or in a bunker that can be emptied by a coupled vacuum hose. The technical problems of such a system appear to be substantially solved, and plans for U.S. field testing are proceeding despite lack of federal interest outside CEQ and EPA. (In September 1976 ERDA's coal technologists did not appear to have heard of Fluidfire or Stal-Laval, and in April 1976 a leading industrial coal expert insisted that Professor Elliott's devices, which the author has seen operating, could not possibly exist.) Dr. John Davidson of CEQ is managing a small project on domestic scale fluidized beds. A paper prepared in connection with it cites an Exxon estimate (ER&E, "Application of Fluidized Bed Combustion to Industrial Boilers," contract CO-04-50168-00, July 1976) which apparently concludes that fluidized beds can capture 100 GW(t), or 36 percent of the industrial fossil energy market (?) by 2000; detailed assessments are clearly needed.

[4] See Chapter 6, note 55.

[5] See Chapter 2, note 28.

[6] See Chapter 2, note 26; and H. Harboe, personal communication, 1976. Pilot tests have given Stal-Laval high confidence of success in avoiding turbine blade erosion by sintered ash, largely because the proposed design does not strive as U.S. schemes do for high turbine inlet temperature. Instead the temperature is limited to about $800°C$, so that sintering, NO_x formation, and volatilization of metallic impurities are greatly reduced. The resulting system, operating at 15-20 bar, attains a First Law turbine efficiency of only 27 percent, but boosts this to over 70 percent (plus any bonus for heat pumps) by taking district heat-

After distribution losses, the Stal-Laval system would deliver 60 MW(e) plus 105 MW(t) of district heating. If we assume that the electricity is used in heat pumps with COP = 2, the total delivered space heat is 225 MW(t). Recent estimates[7] of total construction costs in 1976 dollars, including interest and standby boilers, are M\$40 for the plant and M\$20-25 for the district heating grid[8]; both appear generous.[9] If all electrical distribution were counted as marginal—which it might not be—it would add about M\$35 (see Chapter 6.1). Heat pumps at a more than ample \$200/kW(e) would add M\$12. The grand total implies a system cost of k\$33.4/(bbl

ing off the bottom of the cycle—a concept alien to the electrically oriented U.S. designers. Failure to consider the implications of the Swedish design philosophy appears to account for most U.S. skepticism about the readiness of fluidized bed gas turbines for prompt commercial application. The erosion problem is made still easier by a closed cycle air bypass (see note 10, *infra*, and Harboe and Maude, Chapter Two, note 26). In an encouraging recent development, Stal-Laval has undertaken with Babcock & Wilcox (U.K.) and American Electric Power Company a joint feasibility study for a 64 MW(e) coal-fired fluidized bed gas turbine. If the results are favorable when the study is finished in mid-1977, they intend to build such a plant.

[7] *Ibid.*

[8] This district heating investment of k\$13-16/(bbl · day) is lower than the k\$19.5 computed by Karkheck et al. (see Chapter Two, note 17) because the latter assume an unnecessarily costly method of laying urban pipe. A detailed Swedish study shows that the marginal capital cost of distributing electricity for heating Stockholm buildings is nearly twice that of distributing hot water for the same purpose (considering only the capital costs of the distribution networks). The district heating would be done by tunneling horizontally some thirty meters deep in rock, then tunneling up into basements to connect rather than tearing up the streets. A sixty centimeter pipe on a 120 to 65°C cycle can supply 200 MW(t) (see note 4, *supra*). For details, see "Utredning om Södermalms framtide värmeförsjörning, met förslag till värmeplan för Söder-malm" (Stockholms Kommunstyrelse, 1974), summarized in Neil Muir's informative district-heating paper in the International CIB Symposium (see Chapter Two, note 15). Squires (see note 4, *supra*), states that Enköping, a typical city of 20,000, has 75 percent retrofitted district heating in only six years at a cost below about \$100/kW(t), and that except for Stockholm, which will take until about 2000 to build hundreds of kilometers of tunnels, essentially all Swedish urban areas over 10,000 in population should have district heating by 1980. The pioneering scheme at Västerås took seventeen years to build up the load of its 500 MW(e) + 900 MW(t) station for its population of 100,000, but the Swedes have since devised quicker methods. Now, with the Swedish furnace industry growing at 18 percent/y, retrofits are greatly speeded by a "lending library" of furnaces in which used furnaces can be traded up to successively larger ones as neighborhoods and regions gradually accrete ever larger district heating loads (see note 4, *supra*). Mobile boilers are often used as a transitional tool.

[9] For example, the Thermo-Electron study (see Chapter Two, note 16) estimates that a new conventional industrial gas turbine of this size costs about \$170/kW(e) installed without, or \$230/kW(e) installed with, a waste heat boiler. This implies M\$11.9 or M\$16.1 for a conventional 70 MW(e) gas turbine, leaving a large sum for the fluidized bed, coal-handling system, cyclones, etc. The gas turbine to be used in the Stal-Laval system is off the shelf.

day) delivered heat. To this must be added a few k$ for the fuel cycle and some allowance for capacity factor. The latter, however, would not make much difference, since the capacity factor should be extremely high, and outages would be covered by the standby boilers already allowed for. (Careful design of the electrical and heat systems together—not of the former with the latter as an afterthought—permits excellent seasonal balancing of the two loads as in a conventional combined heat and power station.) The corrections for fuel cycle and capacity factor could probably both be more than covered by conservatisms in the cost estimates. If this is correct, a round number in the vicinity of k$30/(bbl · day) is a realistic whole system cost in 1976 dollars.

It would be useful to have firmer cost estimates for fluidized bed industrial boiler backfits—at first atmospheric, but then at elevated pressures suitable for later conversion to gas turbine cogeneration, via either direct or closed[10] cycles. The sharing of investment between electrical and process steam functions seems attractive[11] : indeed, utility-owned U.S. industrial cogeneration in just three sectors (petroleum refining, chemicals, and pulp and paper) could by 1985 be profitably generating surplus electricity equal to 28–63 percent of all U.S. utility sales in 1974.[12] Better field data for pressurized fluidized beds—generally considered cheaper than those at atmospheric pressure—should become available in the next year.

---

[10] K. Bammert & G. Groschup, "Status Report on Closed-Cycle Power Plants in the Federal Republic of Germany," ASME paper 78-GT-54 for presentation at the Gas Turbine Conference and Products Show, New Orleans, 21–25 March 1976, a highly informative paper kindly called to my notice by Professor R.H. Williams. Skeptics of open cycle coal-fired gas turbines should find closed cycle gas or steam turbines unexceptionable. A 350 kW(e) coal-fired fluidized bed closed cycle gas turbine is being built at Oak Ridge: see quarterly ORNL progress reports by A.P. Fraas on the MIUS project for HUD/ERDA. The eventual 500 kW(e) turbine is to have 26 percent First Law efficiency, up to about 80 percent system efficiency (with waste heat utilization), and a 900°C limestone bed to handle high sulfur coal. The design appears somewhat overelaborate by Swedish standards (see note 6, *supra*).

[11] See Chapter Two, note 16, for three important studies. Dr. R.H. Williams has pointed out that Weyerhaeuser has published an informative, well illustrated prospectus for a cogeneration system (from K.W. Rinard PE, Eugene Water & Electric Board, Eugene, Oregon). It appears that cogeneration could be so attractive as to compete economically with solar heat for both space (with heat pumps) and process, on certain assumptions about allocation of costs between the dual outputs. But cogeneration would rely on depletable fuels and thus would be transitional only. As the use of electricity was gradually trimmed back to appropriate end uses (see Sections 2.5, 4.2, and 8.2), cogeneration would become less attractive as a source of electricity relative to already installed hydroelectric capacity, and as a source of process heat relative to solar process heat, materials recycling, and process redesign (see Chapter Nine, note 34).

[12] S.E. Nydick et al., *op cit.*, Chapter Two, note 16.

Meanwhile, the cost of atmospheric pressure fluidized beds—usually estimated at 10–15 percent less than the cost of conventional combustors—could be used as an upper bound. Desulfurization in the bed by injection of regeneratable limestone or (preferably) dolomite, which can capture about 90 percent of sulfur, is estimated[13] to add less than 1 percent to capital cost and about 3 percent to operating cost—a negligible burden. Costs of pressurized trash and wood burners using fluidized beds (usually deep, not shallow) are also relevant and should be examined: for example, Combustion Power Company's 11 MW(e) + 61 klb/h CPU-400 system (reportedly costing about k$15/(ston · day) input).

7.2. WIND SYSTEMS

All modern data on wind systems refer only to wind-electric systems. It is possible that mechanical work from wind may be used more cost-effectively for pumping heat or water, compressing air, etc., than for generating electricity (see Chapter 2.5), and such applications as water pumping are of urgent interest in many parts of the world. One of the few wind-hydraulic systems under development today is the New Alchemy Institute's Hydro-Wind project, which uses hydraulic coupling to an electrical generator at the Ark on Prince Edward Island.

Wind-electric data are in rapid flux because the field is developing very rapidly. Design philosophies also differ markedly. For example, Lockheed-California, properly relying on mass production cost specialists to examine a mature and familiar kind of engineering, has estimated for ERDA[14] production costs around $610/kW(e) peak and $520/kW(e) peak for 1.7 and 4 MW(e) horizontal axis machines respectively. These would produce busbar electricity at 2.35 and 2.1 ¢/kW-h(e) at average windspeeds of 7 m/s (about 16 mi/h) with an average capacity factor of around 0.5. On the other hand, a New England firm[15] reportedly is planning to market in 1977–1978 a lower technology but reliable and highly sophisticated three-bladed horizontal axis machine of 30 kW(e), priced at $3000 ($100/kW(e) peak!) FOB factory, plus about $1000 for the tower

[13] United Nations Economic & Social Council, Economic Commission for Europe, ENV/R.43, (Geneva: 15 December 1975), p. 30. In his paper for Hartnett (note 3, *supra*), Squires describes British tests that achieved SO_2 capture of about 85–98 percent as the Ca/S atom ratio varied from 1.0 to 2.0.

[14] Lockheed-California, "Wind Energy Mission Analysis: Preliminary Survey for ERDA" (Burbank: 1976).

[15] James Mackenzie, personal communication, December 1976. The name and location of the firm are unfortunately still confidential.

and a smaller sum for installation. Likewise, the frugal two-bladed 2 MW(e) machine at Tvind, Denmark,[16] costs only about k$350, or $175/kW(e) peak, compared with about k$10/kW(e) peak for the NASA Sandusky 100 kW(e) machine now operating, or about k$4.7/kW(e) peak for the 1.5 MW(e) General Electric machine to be commissioned in 1978. In this light, Sørensen's estimate[17] of k$1.33/kW(e) *average* for mass production at MW(e) scale (1974 dollars), some 35 percent more than the 1958 Gedser machine would cost if built today, appears highly conservative[18], i.e., too high for elegantly simple designs.

Thus the available data disagree markedly, depending largely on technical philosophy. Data on some speculative designs, such as James Yen's Grumman vortex tower concept, are not available at all. It may therefore be useful, as a conservative baseline, to consider the costs of a currently commercial 200 kW(e) vertical axis Darrieus device made by Dominion Aluminum Fabricating, Ltd. (Missisauga, Ontario). The first such device, sited on the Magdalen Islands in the Gulf of St. Lawrence, is on the Hydro-Québec grid. It has two blades in tensile stress, is 24.4 m (80 ft) in diameter, and has a 36.6 m (120 ft) shaft atop a 9.2 m (30 ft) tower. It is expected to supply about 0.5 GW-h(e)/y to the grid at 3.5-4.0¢/kW-h(e) from a site with average windspeed of 8.0-8.7 m/s (18–19.5 mi/h). The installed capital cost, turnkey except foundation and grid connection, is k$235 (1976 Canadian dollars, essentially the same as U.S. dollars). But this cost includes virtually the whole investment in the large extrusion dies. Accordingly, the second unit has a list price of only k$175 FOB factory (exclusive of switchgear), with further marked declines thereafter. The run-on production cost in small lots (tens) is probably k$90-100 and the corresponding price about k$130. Cost estimates for larger orders, or for the 1 MW(e) unit under development, are not yet available but are presumably lower.

If we take the total installed price for small-lots production as k$150 (1976 Canadian dollars) and assume a capacity factor of

---

[16] Denmark once had about 100,000 wind machines supplying local electricity, but cheap oil and some powerful institutions put them out of business by the 1930s. See *New Scient.* (10 June 1976), pp. 567 and ff. Similar stories could be told in the Netherlands and in the Great Plains of North America.

[17] B. Sørensen, *Bull. Atom. Scient. 32*:7, 38–45 (September 1976).

[18] Sørensen, *ibid.*, calculates a busbar cost of 1.3 ¢/kW-h(e) without substantial storage, but only 1.7 ¢ with it: not an important difference. In a pioneering wind spectral analysis ("On the fluctuating power generation of large wind energy converters, with and without storage facilities" [København: Niels Bohr Institutet, 1976], summarized in *Science 194*:935-7 [1976]) he shows in considerable detail that ten hours of storage makes a typical Danish wind machine into at least as reliable a source of firm power as a large LWR.

about 0.3, characteristic of most respectable sites, the capital cost is k$2.5 per average kW sent out. Since trunk transmission would be avoided by feeding directly into local lines, distribution losses of about 4 percent seem reasonable. If we assume that a third of the 600-VAC-and-under distribution investment (altogether of order $380/kW) is at the margin, the whole system cost might be about k$204/(bbl · day) delivered enthalpy, less than that of a LWR (see section 6.1). It might be more reasonable to assume instead that the entire distribution investment has already been sunk, however, just as one might assume the same for pipelines associated with a new coal-gas plant. This is because the wind system would normally be used as a fuel saver integrated into an existing fossil-plus-hydro grid (and, eventually, a pure hydro grid). Such supplementary use makes more sense transitionally than an attempt to imitate, with full storage[19], the operation of existing central stations. (This view is consistent with Chapter Two's approach to end-use matching—not, of course, with absurd proposals [see Chapter Five, note 39] to run aluminum smelters with wind machines.) It might make even more sense to use the mechanical work directly to pump heat or to compress air for storage at or near the point of end use, since such storage is simple and cheap (see Chapter 2.5). (Also at high latitudes the wind blows mainly in the winter when heat needs are greatest and sunlight is scarcest.) But those concerned to attempt an irrational imitation of reactors with wind machines will find the costs of doing so surprisingly low.[20]

7.3. GEOPHYSICAL AND BIOCONVERSION SYSTEMS

Reliable cost estimates are not yet—but may soon be—available for wavepower technology, being vigorously and impressively developed (chiefly by Stephen Salter at Edinburgh University) with U.K. government funding. The technology does appear, though, to be competitive with other U.K. electricity sources, and is capable of yielding large amounts of energy. Production from about 1000 km or less of collectors would exceed all present U.K. electrical demands (only a small part of which are for appropriate uses), as the annual average power of waves off the western U.K. is typically 25–75 kW/m, most of which can be captured.

Another interesting technology—also not considered in Chapter

[19] *Ibid.*
[20] *Ibid.*

Two—is small hydroelectric sets, both dammed and run of the river, similar to those that used to be common in northern New England before they were bought and shut down by utilities. Hayes[21] estimates that the People's Republic of China had over 60,000 such machines in late 1975, ranging from 0.6 to about 100 kW(e) and totaling 2.5-3 GW(e). Small hydro sets produce most of the electricity for half the production brigades and over 70 percent of the communes in China. Several visitors have reported large (in a few cases 200-300 kW[e]) Chinese hydro sets, but the general pattern is small sets with 0-5 m heads, including 5 kW run of the river wooden turbines. A National Academy of Sciences panel[22] has also reported wooden turbines in the Soviet Union.

The same panel has called attention to the hydraulic air compressor, or trompe, an exceptionally simple, cheap, and effective device with no moving parts. It uses hydraulic potential to compress air. For example, a 1907 report of a twenty-two meter fall of water through a one and one-half meter pipe claimed that air was compressed to 8 bar with an output of about 750 kW at 82 percent efficiency.[23] A working model is in a Toronto science museum.

Cost estimates will not be developed here for geothermal heating (a more attractive system than geothermal-electric), partly because costs are very site-dependent and partly because geothermal sources, being depletable, are not truly soft (and not always pleasant). It is interesting, however, that technical advances in telethermics now permit very remote siting of wells from heat loads. For example, an insulated pipe in Italy is reported[24] to have shipped hot water continuously since 1969 for over one hundred kilometers at an average thermal efficiency of 98.5 percent.

The most authoritative sources (including the U.S. National Research Council's CONAES study) estimate that bioconversion of crop residues and municipal wastes will yield fuels costing less than $2/GJ (an attractive $11.60/bbl). Estimates of capital cost vary and depend on the method assumed. One consistent set of estimates[25] in 1974 dollars suggested values from less than k$10/(bbl · day) to about k$20/(bbl · day) for crop residues, and up to as much as k$30—often less—for municipal waste pyrolysis (which can offer

[21] D. Hayes (Washington, D.C.: Worldwatch Institute), *Rays of Hope* (New York: W.W. Norton, 1977). See also Chapter Five, note 28.

[22] U.S. National Academy of Sciences, *Energy for Rural Development* (Washington, D.C.: NAS, 1976).

[23] *Ibid.*

[24] A.E. Haseler, personal communication, 1976.

[25] FEA, Solar Task Force Report, *Project Independence Blueprint* (Washington, D.C.: USGPO 4118-00012, November 1974).

substantial economic credits for recovered materials and for saved disposal costs[26]). For most agricultural projects, k$13-20/(bbl · day) (1976 dollars), plus a few k$ for local delivery systems, might be typical, including collection investment.[27] On the other hand, the local distribution might use existing equipment. Some data on collection costs are available—for example, in Finland and Sweden, which obtain respectively 14 percent and 7 percent of their total energy budgets from wood (including wood wastes).[28] The cost, efficiency, and diversity of conversion systems promise to improve rapidly through sophisticated bacterial and enzymatic methods now under study,[29] though Chapter Two assumes no such systems. More traditional methods such as anaerobic digestion[30] have already arrived in the U.S. on a substantial scale, for example in the 23 Mm^3/y (820 × $10^6 ft^3/y$) Oklahoma City feedlot converter of Calorific Recovery by Anaerobic Processes, Inc., approved by the FPC in May 1976 as a source of pipeline gas.

[26] Good data should emerge from the twenty-two projects summarized in *Recommendations for a Synthetic Fuels Commercialization Program* (Synfuels Interagency Task Force report to President's Energy Resources Council, November 1975), vol. III, p. III-D-35.

[27] A.D. Poole & R.H. Williams, *Bull. Atom. Scient. 32*:5, 48-58 (May 1976), and its citations, gives probably the best available summary; M.J. Antal, Jr., *ibid.* pp. 59-62; J. Marshall et al., E-X-25, Forestry Service, Environment Canada, February 1975; InterGroup Consulting Economists Ltd. (Winnipeg), "Economic Pre-Feasibility Study: Large-Scale Methanol Fuel Production from Surplus Canadian Forest Biomass" (Ottawa: Policy and Program Development Directorate, Environmental Management Service, Fisheries and Environment Canada, September 1976); Forest Service, USDA, NSF/RA-760013 (March 1976), inferior to the Canadian papers; MITRE Corporation (McLean, Virginia), "Bioconversion (Fuels from Biomass) in the United States" (June 1976) and "Silviculture Biomass Plantations" (February 1977); W.E. Scott, *Energy International 13*:7, 19-22 (1976), on an Australian proposal; C.G. Golueke & P.H. McGauhey, *Ann. Rev. Energy 1*:257-77 (1976); K. Sarkanen, *Science 191*:773-76 (1976), and adjacent papers; D. Hayes, "Energy: The Case for Conservation," Paper 4 (Washington, D.C.: Worldwatch Institute, 1976); *Capturing the Sun Through Bioconversion* (March 1976 conference proceedings, The Washington Center [1717 Mass. Ave. NW, Washington, D.C. 20036]); H.G. Schlegel & J. Barnea, eds., *Microbial Energy Conversion*, UNITAR seminar, Göttingen, 4-8 October 1976 (Göttingen: Erich Goltze KG, 1976); NAS, *supra* note 22; Chapter Five, note 29; *Economist*, 26 February 1977, p. 85.

[28] *Supra* note 21.

[29] See e.g. L.A. Williams, "*Dunaliella*: Growth and Glycerol Production" (Stockholm: Bakteriologisk Biotechnik, Karolinska Institutet, 1976); or, for a broader review, R.E. Anderson, "Food and Fuel Self-Sufficiency Through Modern Biology," in *Biosciences: A Challenge to the Individual and to Society —Socio-economic and Ethical Implications of Enzyme Engineering* (New York: UNESCO, 1977). *Cf.* C.W. Lewis, "Fuel Production from Biomass" (London: International Institute for Environment & Development, 1977), whose generally unfavorable net energy yields appear to result from the use of inappropriate technologies for heating, stirring, etc.

[30] See Chapter Five, note 29.

7.4. SOLAR HEAT SYSTEMS

Solar process heat,[31] though attractive and starting to be widely studied, is not yet quite in a state where definitive cost estimates are generally accepted for a wide range of designs. Useful numbers are about to emerge from some studies for ERDA,[32] though their methodology and technical range in this task are somewhat constrained. Low temperature process heat is straightforward today with air or water heated in flat plates priced at $65/m² or less, and the heat at that collector price costs of order $3.50/GJ or less in the U.S., equivalent to $14.50/bbl oil burned at 70 percent efficiency in a free device.

High working temperatures can be obtained not only by various forms of concentrators, tracking or fixed (it is possible to make wide aperture nontracking concentrators), but also by highly selective surfaces, either on flat nonfocusing plates or in cylindrical configurations that concentrate severalfold. For example, thin film sputtering technology now suffices to produce selectivities (ratio of visible absorptivity to infrared emissivity) of 50-60+,[33] at pilot scale costs of a few $/m². It can be readily calculated that such a surface, suitably insulated and contained in a hard vacuum, is so insensitive to cloudiness that it should provide working temperatures, under load, of order 400-600°C on a cloudy day in the winter at Scandinavian latitudes. Obviously there is also room for technical simplicity in sunnier places: one type of effective solar cooker 1.4 meters across, for example, now sells for $6.70 in India.[34]

[31] E.P. Gyftopoulos et al., *Potential Fuel Effectiveness in Industry* (Cambridge, Massachusetts: Ballinger for The Energy Policy Project, 1974); J.A. Day et al., UCRL-76390, May 1975; D.P. Grimmer & K.C. Herr, LA-6597-MS, December 1976, which quotes very low installed prices ($118/m²) for 9X-concentrating Winston collectors working at 315°C; and a large international literature. Solar crop drying seems ready for immediate large-scale use: see, for example, the experiments at Georgia Tech and the wide international experience cited by NAS, note 22, *supra*. Solar distillation for desalination is in widespread use at semi-industrial scale (up to 8600 m²) in many countries. A further high quality solar application—lighting—should not be neglected: W.D. Metz, *Science 194*: 1404 (1976).

[32] See Chapter Four, note 4.

[33] A.B. Meinel, in "Briefings before the Task Force on Energy," Subcommittee on Science, Research, and Development, Committee on Science and Astronautics, USHR, Serial U, 92d Cong. (Washington, D.C.: USGPO, 1972), and in "Solar Energy for the Terrestrial Generation of Electricity," hearing before the Subcommittee on Energy, Committee on Science and Astronautics, #12, 93d Cong. (Washington, D.C.: USGPO, 1973). The α/ϵ of about fifty which Meinel mentions, with an apparent lifetime of order forty years on the basis of accelerated lifetime tests, has since been bettered in proprietary work. Wet chemical deposition at much lower cost can reportedly yield films with α/ϵ at least thirty, but their lifetime is not yet known.

[34] *Supra* note 21.

The long-run marginal cost pricing philosophy (or its functional equivalent in subsidies or in how investment decisions are made symmetrically—see Chapter 3.2) implies a very different approach to solar space heating than that treated in the traditional literature. Seasonal heat storage eliminates backup requirements—and associated potential for utility load management problems—and can increase collector capacity factors and operating efficiency. Long-run marginal cost pricing implies a new kind of optimization of storage volume, collector area, operating parameters, and building design,[35] and must not be misjudged in the light of different design philosophies that attempt to make solar heat compete with cheap fuels. (Indeed, the classical papers on solar space heating, suggesting that solar heat would be too costly to provide more than a third to a half of heat requirements, were even computed at pre-1973 oil prices, yet are still quoted as gospel today.)

The technology of seasonal heat storage is critical to solar space heating, and is more advanced than is commonly realized. Storage of latent heat[36] is elegant but difficult. This chapter therefore assumes the simple, proven technology of sensible heat storage as hot water (which is more difficult and expensive than hot rocks, the equivalent used in hot air solar heating systems). With modern water tank technology—prefabricated tongue and groove concrete slabs assembled modularly in a hole in the ground, then sprayed with insulating foam, lined, and backfilled—the installed price of seasonal hot water storage is about $21-34/m³ (60-95 ¢/ft³) (1975 dollars),[37] with the lower prices characteristic of tanks nearer 100 than

---

[35] See Chapter One, note 26.

[36] W.A. Shurcliff, *Bull. Atom. Scient.* 32:2, 30-40 (February 1976), mentions success by the perservering pioneer M. Telkes, though the details are not yet clear. The latent heat of fusion of water—a much simpler system—is of course used in the ACES system (see note 37, *infra*).

[37] H.C. Fischer, "The Annual Cycle Energy System" (Oak Ridge, Tennessee: Energy Division, ORNL, 1975). An illustrated but less detailed account appears in *PE Magazine*, June 1976. The ACES system pumps heat back and forth between a building and a large tank in which water is frozen during the winter and melted during the summer. The heat pump can attain quite high efficiencies. Ideally, the heating and cooling loads should be balanced; if there is a heat deficit, a solar booster can be used to advantage. Hayes (*supra* note 21) states that the ACES approach was first used in Japan in the 1940s. The U.S. Veterans Administration is building at Wilmington, Delaware, a sixty-bed nursing home with an ACES system: 2000 m³ storage, 200 m² solar panel, maximum heating load 230 kW, maximum cooling load 178.5 kW, peak summer electric load for the heat pump 3 kW, energy savings about 50 percent, and economically attractive by orthodox criteria (see ORNL-5124, April 1976, p. 5). ACES systems need not, in principle, be electrified: they might use fossil-fueled heat pumps (see Chapter Three, note 28, or the miniature MIUS-like systems [see note 10, *supra*] being developed by Fiat and The Open University) and their waste heat.

20 m^3. (Ferrocement tanks sprayed with foam insulation might be even cheaper, since bare 100 m^3 tanks of this type are being built in many countries at 1976 prices of only \$5-10/m^3.) With the more costly concrete-slab design assumed here, the marginal capital cost of the solar heating system now operating with no backup at Lyngby, Denmark,[38] would have been about k\$6 rather than the actual k\$8—or even less without the high costs of first of a kind design.[39] Storage costs will probably decline further as tanks are integrated into buildings: apparently[40] heat sink tanks up to nearly 5000 m^3 are used in Japan, partly as seismic stabilizers.

Estimates of the capital cost of solar space heating systems depend sensitively on both building and collector design. For new buildings, passive systems[41] with negative, zero, or negligible marginal capital costs should suffice. The following discussion thus refers only to retrofits requiring hardware. Furthermore, impressive economies are available in principle from mass production of very simple collectors. Suitable concepts include Dr. Jerry Plunkett's robust paper and phenolic composite honeycomb sandwich that unrolls like roofing paper, replaces shingles, and is estimated to have a very long lifetime (many decades) and an installed cost of a few \$/ft^2; perhaps the Philips tubular collector designed for cloudy northern European latitudes and expected to enter advanced pilot production around 1977-1978[42]; grids of flexible plastic tubing laid in poured asphalt

[38] T.V. Esbensen & V. Korsgaard, *Solar En. 19*:195-199 (1977), and T.V. Esbensen, personal communication, 1976. Seasonal heat storage has also been demonstrated in Canada: see A.B. Lovins, Chapter Two, note 18. For a thorough discussion, see A.B. Lovins, letter to H.A. Bethe, 17 March 1977, to be published in main volume to the U.S. Senate Small Business/Interior Committee hearing, "Alternative Long-Range Energy Strategies" (Washington, D.C.: USGPO, in press, 1977).

[39] In contrast, the marginal cost of all conservation and solar measures on a 57 m^2 house in New Mexico using a 25-30 percent solar system is about \$2300, and the price of gas for backup heating (70-75 percent of a heat load greatly reduced by efficient Arkansas style design—Owens-Corning-Fiberglas, cited in Chapter Two, note 15) is less than \$100/y: *Contractor*, 15 August 1976.

[40] *Supra* note 37. The Ontario Hydro building in Toronto stores 6000 m^3.

[41] Passive systems do not require forced circulation of a working fluid to distribute their heat, but rely on convection, conduction, and radiation. They capture heat not through conventional collectors but through the ingenious use of windows, walls, etc., and the use of the "free heat" provided by occupants, lights, and appliances. Many passive designs are extraordinarily simple and effective in a wide range of climates. See *Proc. Passive Solar Heating & Cooling Conf. & Workshop* (Albuquerque, 18-20 May 1976), NTIS, 1976. Passive systems are hard, though not impossible, to retrofit.

[42] The envelopes used in current Philips tests use low pressure sodium vapor lamp technology and are thus somewhat larger and heavier than fluorescent lamp tubes. The latter, at less than 10¢ each, are perhaps the cheapest manufactured commodity per unit mass, and are extruded at about 13 m/s. Owens-Illinois uses a similar approach. Simpler, cheaper designs, including all-plastic extrusions and plastic-glass hybrids, are being developed.

on a flat roof; plastic tubes[43] imbedded in a lightweight rollable mat like the "Sunmat" made by Calmac[44]; the Thomason system[45]; and the van Arx solar pond[46]. Commercial versions of the Calmac and Thomason systems sell today for about $3-5/ft² installed. Their striking simplicity merits close attention.[47]

To be pessimistic, however, we can assume that solar heating is to be done for buildings needing active hardware—i.e., for existing buildings—by conventional flat plate assemblies that are not integrated into the building structure, have high transport and installation costs, and involve the inherent costs of assembly and materials for high quality glazing and sheetmetal work. In late 1974, factory prices for such collectors were $10-50/m², while various versions assembled and delivered, including moderately selective plates, were $70-120/m² for the collector assembly only.[48] Many solar hardware

[43] In a similarly innovative approach, Professor V. Silvestrini (Institute of Physics, Faculty of Engineering, University of Naples) is experimenting with a selective radiative surface—an aluminized polymer film—that passively cools things it covers by radiating preferentially at infrared wavelengths at which dry air is nearly transparent, while reflecting longer and shorter wavelengths. In one test, an insulated box whose lid was covered with such a film attained temperatures 15°C below ambient. The film has, of course, very low power density, but its cheapness may lend it to agricultural storage applications. See B. Bartoli et al. (also of the Institute of Physics at the University of Naples), "Nocturnal and Diurnal Performances of Selective Radiators," January 1977, submitted to *Applied Energy*.

[44] Calmac Corp., Box 710, Englewood, New Jersey 07631.

[45] Thomason Solar Homes, Inc., 6802 Walker Mill Rd. SE, Washington, D.C. 20027. Several other makers apparently offer similarly simple designs. For a useful survey, see W.A. Shurcliff, *supra* note 36, and his index cited there; the publications of Total Environmental Action, Inc. (Harrisville, New Hampshire); and the usefully eclectic and comprehensive "Energy R&D and Small Business," a six volume compendium issued by the Select Committee on Small Business, US Senate, 1975. See also OTA, Chapter Eight, note 29.

[46] W. van Arx of the Woods Hole Oceanographic Institute (Falmouth, Massachusetts) has extended Tabor's work on solar ponds to higher latitudes and colder climates. A current experimental pond consists of an insulated hole in the ground, 1 m deep and 5 m across, with a layer of coal (as a black absorber) in the bottom. The hole is filled with 45 percent calcium chloride brine covered with 15 cm of fresh water—and nothing else. It sits in the open, acting as a fisheye lens that converges direct and diffuse radiation from the whole hemisphere onto the coal. A few loops of pipe in the brine act as a heat exchanger, yielding about 2.7 kW average heat output at very high temperatures—probably over 90°C. Heat storage in the pond and adjacent soil totals about 4-8 GJ, or about a barrel of oil equivalent. As with Tabor's Israeli solar ponds, the thermal, density, and refractive index gradients in the brine are self-stabilizing. Wind mixing with the fresh water layer is apparently not a problem—especially since periodic freezing at the surface reextrudes the $CaCl_2$ back into the brine.

[47] This message, which emerges clearly from decades of solar experience (see note 45, *supra*), is still alien to most ERDA solar contractors.

[48] J.M. Weingart, "Solar Energy Conversion and the Federal Republic of Germany—Some Systems Considerations," draft WP-75-158 (Laxenburg, Austria: IIASA, December 1975). IIASA studies of solar heating costs—unfortunately without a long-run marginal cost pricing philosophy—are in progress,

production experts estimate that an installed square meter of high quality flat plate collector plus 1 m^3 of seasonal water storage,[49] installed complete with all plumbing, will cost about $150 (1976 dollars) by 1978–79 and about $100 by the mid-1980s (1976 dollars)—the time of interest for comparison with hard technologies ordered now, since the latter have about a ten-year lead time, and we are therefore deciding now whether to build them. In conventional U.S. units, $100/(m^2 + m^3) corresponds to $9/ft^2 in 1985 including seasonal storage; ERDA seeks $10/ft^2 in 1980 without seasonal storage, which costs about $1 per square foot of collector (if storage costs $21–27/m^3 water and requires 0.4–0.5 m^3/m^2 collector, a typical U.S. value). A common installed system price in California today is about $10/ft^2 with slight storage.

Let us apply this $100/(m^2 + m^3) figure—consistent with Office of Technology Assessment and A.D. Little estimates and with detailed analyses by private mass production specialists—to Denmark. A Danish south wall receives average total insolation[50] of about 125 W/m^2. Modern (double-glazed, slightly selective) flat plates in Danish conditions can readily achieve an average First Law efficiency of about 0.42, counting their own heat losses but not those of the rest of the system. Seasonal fluctuations in Denmark do not appear to require—taking account of heat losses in storage and plumbing—more than about 0.7 m^3 of storage per m^2 of collector in a well-insulated house (heat load averaging below 1 KW[t]),[51] thus saving about 8 percent of the estimated system cost. (Actually the collector area and storage volume can be traded off against each other, the optimum depending in detail on climate and building design.) Completely solar space heating for a typical Danish house (125 m^2 floor) in the mid-1980s should thus cost of order k$118/(bbl · day)

including surveys of U.S. and European hardware (mainly of the more complex kinds). IIASA's N. Weyss has devised an ingenious set of graphs for comparing collector areas, storage volumes, and other parameters, assuming a design philosophy consistent with trying to compete with cheap fossil fuels. For surveys of relatively complex hardware in the U.S., see also S.W. Herman & J.S. Cannon, *Energy Futures: Industry and the New Technologies* (New York: INFORM, 1976; and Cambridge, Massachusetts: Ballinger, 1977).

[49] A 100 m^3 tank cycling over 50°C and discharging over a period of four months produces 2 kW(t). The long-run marginal cost pricing philosophy described in Chapters One through Three makes it unnecessary to consider partial solar heating with backup, or combined solar heating and cooling, as tricks to make solar heating economic: the marginal cost of seasonal heat storage is less than that of backup energy supply capacity, essentially because water or rocks cost less than copper, steel, and concrete. Solar cooling is not of much long-term interest, since the need for it can be removed by rational architecture, even in the hottest tropical climates, as millenia of traditional architecture have demonstrated worldwide.

[50] B. Sørensen, *Science 189*:255–60 (1975).

[51] *Ibid.*, and Lovins letter to Bethe, note 38, *supra*.

delivered heat—about k$3.5 for an average heat load of 2 kW (good insulation) or k$7 for a typical 1974 heat load of 4 kW,[52] though in the latter sort of house heavy insulation would be the first priority.

The less favorable regions of the U.S. receive about 180 W/m^2 on an optimally oriented fixed flat plate—about 270 W/m^2 in southern deserts, close to the 290 W/m^2 of much of India. Seasonal heat storage need not be as large in the U.S. as in Denmark because the seasonal fluctuations are smaller. Typical mid-1980s specific investment for completely solar space heating should thus be of order k$50-70/(bbl · day) in the U.S., assuming total insolation of 180 W/m^2. (The extra installation costs of retrofit should be roughly balanced by savings through the reuse of existing hot water plumbing or hot air ductwork, though more detailed study of this point would be useful.) All such figures should be treated with caution, not only because they assume fixed technologies in a rapidly advancing field, but also because, as Weingart points out,[53] the installed price even of a completely conventional, nonsolar domestic water heating system varied by a factor of *more than two* in a recent Southern California survey of similar apartment buildings. This variation arose from the variable local costs of labor and materials and the variable skill (and avarice) of contractors.

The few integrated solar heating and cooling projects now built or being built on a large scale are probably also a poor guide to future costs. For example, the Los Alamos project,[54] which ran a $150/$m^2$ collector cost up to a $690/$m^2$ installed collector cost, appears to be a typical example, like many ERDA funded solar demonstration projects, of a highly sophisticated and instrumented, therefore excessively complex and costly, installation.[55] Simpler systems, such as the Oss houses[56] (where two-thirds solar heating competes even with cheap Groningen gas in a cloudy high latitude country), may be a better measure of realistic overhead levels.

[52] *Ibid.*, and *supra* note 38. Seasonal storage is hard with such poor insulation.
[53] *Supra* note 48.
[54] *Ibid.*
[55] The major aerospace and war corporations heavily involved in the ERDA solar program have useful technical resources, but may not be able to keep the hardware simple enough and the overheads low enough. (Just *responding* to a recent ERDA coal RFP would have cost M$2!) There is a moral in the story of the new Boeing-Vertol subway cars for Boston. Their first design reportedly used 1300 parts per door. This was reduced, with difficulty, to about 300; the doors may now work, but clearly the designers have become so sophisticated that they can't design a door any more. Mass production is not an answer to this institutional problem (see Herman and Cannon, note 48, *supra*).
[56] H. van Bremen & J.M. van Heel, "Solar Energy in Local Authority One-Family Houses in Holland," International CIB Symposium (see Chapter Two, note 15). Rapid developments in European solar architecture are to be documented by a new newsletter ("Euro-Sol Bulletin") and other publications from Mirtech SA (Chemin de la Dauphine, CH-1299 Commugny, Switzerland).

Chapter 8

Comparative Capital Costs and the Role of Electrification

8.1. COMPARATIVE CAPITAL COSTS

The computations in Chapters Six and Seven are summarized in Table 8-1, subject to the many conditions and caveats given in Chapters Three (section 3.2), Six, and Seven. All the figures neglect future cost escalation beyond the GNP inflation rate, consider only those technologies that are already demonstrated (except for coal synthetics and, arguably, some other fossil fuel technologies and large-scale fluidized bed gas turbines), and take no account of the quality of energy delivered—a point discussed in detail below.

Even if one quibbles with details, the general shape of Table 8-1 suggests several broad conclusions already used in Chapter Two:

1. big electrical technologies are pricing themselves out of the capital market[1];
2. improved end use efficiency[2] will long remain the most cost-effective investment;

---

[1] It is no wonder that the self-financing ratio of U.S. utilities fell from 0.40 to 0.19 during 1960–1973 (D.P. Kamat et al., "Regulatory and Tax Alternatives and the Financing of Electricity Supply," Research Report #7 [Austin: Center for Energy Studies, University of Texas, NSF-RA-N-75-123, September 1975], p. 29), nor that institutional investment in utility debt has fallen off sharply in the past few years. The Invisible Fist strikes again!

[2] Some remarkable examples can be found in such mundane devices as domestic furnaces. C.R. Montgomery, president of Michigan Consolidated Gas Co., has described a $35 gas furnace retrofit which, in the Detroit area, saves 26.5 percent of the gas—a specific investment at least an order of magnitude below that of

Table 8-1. Approximate Marginal Capital Investment (1976 dollars) Needed to Build Complete Energy Systems to Deliver Energy to U.S. Consumers at a Rate Equivalent to One Barrel of Oil Per Day (About 67 Kilowatts) on a Heat-Supplied Basis (Enthalpic, Without Regard to Quality of Energy Supplied)

Energy System	$/(bbl · day)	Form Supplied
Traditional direct fossil fuels, 1950s–1960s; or direct U.S. coal, 1970s	2-3,000	fuel
North Sea oil, late 1970s	10,000	fuel
Frontier oil and gas, 1980s	10-25,000	fuel
Synthetic fuels from coal or from unconventional hydrocarbons, 1980s	20-40-70,000	fuel
Conventional central coal-electric with scrubbers, 1980s	170,000	electricity
Nuclear-electric (LWR), mid-1980s	200-300,000	electricity
"Technical fixes" to improve end-use efficiency:		
new commercial buildings	-3,000[a]	heat+
common industrial and architectural leak plugging	0-5,000[a]	heat+
most heat recovery systems	5-15,000[a]	heat
worst case, very thorough building retrofits	25,000[a]	heat
Coal-fired fluidized bed gas turbine with district heating grid and heat pumps (COP = 2), early 1980s	30,000[a]	heat
Retrofitted 100 percent solar space heat, mid-1980s, with no backup required, assuming costly traditional flat plate collectors and seasonal storage	50-70,000[a]	heat
Bioconversion of agricultural and forestry wastes to fuel alcohols, around 1980	13-20,000	fuel
Pyrolysis of municipal wastes, late 1970s	30,000[b]	fuel
Vertical axis 200 kW wind-electric, late 1970s	200,000	electricity

[a]These costs include the cost of end-use devices, which are often very expensive. An unpublished 1976 Shell analysis calculates typical capital requirements, in ~1976 dollars per primary bbl/day *used*, of order k$120-200 for a European car, k$35 for a conventional house heating system, k$5-10 for industrial boilers, and k$14 for a blast furnace.

[b]Excludes investment credit for byproducts (e.g., materials recovery) and for waste disposal services replaced.

3. though all major domestic supply technologies available at the margin (except direct coal) are relatively capital-intensive, the life cycle and even the capital costs of soft and transitional technologies, calculated traditionally as internal costs per unit of energy

Arctic gas supply. For average fossil fuels, moreover, saving one end use unit now saves about 2.4 units of fuel resource in the ground—a figure expected to increase with increasing electrification and fuel synthesis (United Nations Economic Commission for Europe, "Increased Energy Economy and Efficiency in the ECE Region," E/ECE/883/Rev. 1 [Geneva: ECE, and New York: UN, 1976], pp. 72-73).

or power, are arguably attractive compared with those of competing complete hard technology systems;
4. but to make such comparisons specific, one must examine end
uses one at a time.

To take up this last point, consider the task of supplying heat at
modest temperatures. Chapter Four shows that this is the dominant
term (35 percent) of end use in the U.S. and an even larger term elsewhere—typically over half of all end use in Europe. Chapter 7.4
showed that completely solar space heating in an existing mid-1980s
U.S. building (assuemd to be very well insulated, which is worthwhile
no matter how it is heated), using relatively costly technologies within the present art, will have a whole system marginal capital cost of
about k$50–70/(bbl · day) delivered heat. (The corresponding figure
for Denmark would be about k$118/[bbl · day]). Let us suppose
that the long-run alternative, to which we are being asked to commit
resources today, is a LWR-powered electric grid, shown in Chapter
6.1 to cost k$211 or k$235/(bbl · day), operating a heat pump costing, say $200/kW(e) and having a COP = 2.5. (A calculation for nu-
clear district heating would of course yield a very similar result.) The
whole system marginal capital cost of the nuclear electricity and heat
pump system is thus

$$\frac{\text{k\$211/(bbl} \cdot \text{day)}}{2.5} + \frac{(\text{\$200/kW(e)}) \times (67.1 \text{ kW(t)/(bbl} \cdot \text{day))}}{2.5 \text{ kW(t)/kW(e)}}$$

or k$90/(bbl · day) delivered heat, assuming the lower nuclear cost
derived from the simple regression 1974–1976 escalation of LWR
cost. Assuming instead the k$235/(bbl · day) nuclear cost (see Chap-
ter 6.1) would make the output of the nuclear-powered heat pump
cost not k$90 but k$99/(bbl · day). Both these figures are substantially higher than the solar heating capital cost for the U.S., k$50–70,
and nearly comparable to that for Denmark, k$118/bbl · day).

This comparison favors the nuclear alternative by omitting its fuel
cycle costs[3] and many uncertain but unquestionably real terms of
nuclear system cost (see Chapter 6.1). Together these probably
amount to a factor of two over the nuclear cost assumed here. This
comparison also assumes solar hardware more complex than necessary and ignores neighborhood scale solar systems or those integrated

---

[3] See Chapter Three, note 26. Completely solar heat, even in Denmark, is also
cheaper than heat from a furnace fueled with synthetic fuels, if lifetime fuel costs
(even discounted) are taken into account.

with building structure, all of which would reduce solar costs. The externalities of both nuclear and solar systems—a better basis for decision than the internal costs—are omitted. Taking all these factors or even the internal costs alone into account, the economic advantage of the solar over the nuclear system, even under Danish conditions, seems so robust as to survive any uncertainties in the cost calculations.

Numerical quibbles apart, several methodological objections might be raised to this calculation. First, it takes no account of user diversity—for the cogent reason set out on pp. 112–113 in Chapter 6.1. Second, it takes no account of the likelihood that part of the winter peak electrical load will in fact be met by lower capital cost generation such as gas turbines. This point is considered in the same place. Third, it does not consider the attractions of, say, an ACES system[4] with the potential for higher effective COP. True; but neither does it come close to a true and full accounting of the nuclear costs or of the scope for reducing the solar costs, and it is hard to imagine a COP high enough to make up the difference.

This sort of argument, which can be reproduced for each other end–use thermodynamic category and for each group of competing hard and soft technologies, is complex and perhaps irresolvable. It does, however, leave an impression that soft system economics leave a strong prima facie case to be answered by those proponents of hard technologies who consider narrow economic rationality more important than all the external costs considered elsewhere in this book.[5]

8.2. THE ROLE OF ELECTRIFICATION

There are broader perspectives from which to view the kind of com-

---

[4] See Chapter Seven, note 37.

[5] An equally strong prima facie case today can be found in the cost comparison between electricity and direct fuels. For example, if fuel oil in the mid-1980s cost, say, $30/bbl (1984 dollars), while nuclear electricity cost 5 ¢/kW-h(e) (1984 dollars) at the busbar and only 8 ¢/kW-h(e) delivered, then an existing (free) oil furnace of 67 percent First Law efficiency would compete with an existing (free) heat pump with COP not over 2.87. If capital costs of the end–use devices were not treated as sunk, this limiting COP would probably rise. It is no wonder that Table 4-1 shows no electrical water heating or cooking or space heating, and only partial electric air conditioning, in the cost-conscious commercial sector, despite electrical rate structures that favor large users. Likewise, an October 1976 draft staff study by the California Energy Resources Conservation and Development Commission ("Potential Impact of Natural Gas Shortages on California Electricity Demand," Office of Economic & Data Analysis, Energy Assessment Division) suggests that, largely because electricity is so costly, demand for it as a substitute for gas is unlikely to be significant during projected curtailments of gas supply.

parison just performed in section 8.1. Consider, for example, First Law efficiency. The First Law efficiency of a soft technology like a solar system is not an interesting criterion: not only for the reasons adduced by Shurcliff,[6] but also because what is "lost" (not captured and delivered) is renewable, not depletable,[7] and is either desirable for albedo compensation or (in other soft systems) usable locally as low grade heat because of small unit scale. The only reason for seeking to increase the First Law efficiency of, say, a solar collector is to obtain more heat per dollar invested—not as an end in itself, to prevent the depletion of a dwindling resource. But the First Law efficiency of hard technology electrical systems is of great interest. In Britain, for example, including the fuel cycle and the grid losses, it is about 0.25. This means that four units of high grade fuel are used to produce one unit of delivered enthalpy. That delivered enthalpy is indeed of low entropy (high quality); but to compensate for the low First Law efficiency of the electric system would require the equivalent of a reciprocal COP, say about three to four, in *each* significant electrical end use, in order not, in a real and irreversible sense, to waste precious fuel compared to that needed to perform the same task directly.[8] Analogously, in what major applications does electricity have a high enough First Law efficiency of end use to compensate for its order of magnitude penalty in capital intensity (see Table 8-1) relative to many competing energy systems?

Second Law efficiency, a more subtle and useful criterion,[9] makes the same argument emerge even more strongly. Recall that the Second Law efficiency of a typical U.S. domestic gas furnace system in winter is of order 0.05, that of a typical New York City office air-conditioning system of order 0.02, that of a car of order 0.12, and that of a directly fueled industrial furnace of order 0.30.[10] The reason that the first two figures are so much lower than the third and fourth is plain: they are low grade functions being met in a way that entails thermodynamic overkill. Any process that uses a flame temperature of kilodegrees or a nuclear reaction temperature of teradegrees to produce a steam temperature of centadegrees to drive

---

[6] See Chapter Seven, note 36.

[7] Thus the definition of "primary energy" shifts during the transition from hard to soft technologies. For soft technologies it shows not the natural energy flux incident on the collecting device, but rather the useful output. This approach is used in, e.g., Figure 2-2.

[8] *Supra* note 5.

[9] American Institute of Physics, Conference Proceedings #25, *Efficient Use of Energy* (New York: AIP, 1975); E.P. Gyftopoulos, see Chapter Seven, note 31; M. Ross & R.H. Williams, see Chapter Two, note 12.

[10] *Ibid.*

turbines and make electricity of infinite effective temperature, all to make an end-use ΔT of dekadegrees, is going to have an abysmal Second Law efficiency that heat pumps cannot greatly alter.[11] Worse, such an approach is inelegant—a cardinal sin to a physicist. Natural energy fluxes, on the other hand, are of relatively low thermodynamic performance when used diffusely (luckily for fragile organisms like us) while having higher potential that need not be used if not needed. The $5500° K$ potential of sunlight, the kilodegree potential of fixed carbon, and the infinite temperature of wind- or water-derived mechanical work need only be concentrated for applications requiring them,[12] thus preserving high Second Law efficiency through end-use matching.

To these engineering arguments must be added the sociopolitical arguments developed in Chapters 2.9, 5.5, 9, and 10. These issues appear in both political and moral terms to outweigh any conventional engineering or economic considerations of the marginal sort adduced in this chapter (see Chapter 3.2). But since the comparative social costs of energy systems are set out elsewhere, they will not be listed here again.

Are there not countervailing arguments? Of course there are, notably the convenience with which a user can control electricity.[13] But convenience, notoriously hard to measure, is not necessarily as important a criterion as some have proposed, especially at the margin. (The next kW-h(e) improves one's life far less than the first one did.) Electricity is not even indispensable for many of its currently common and appropriate uses. In my own experience, gas mantle lights, for example, are not significantly more trouble to use and

[11] Even in France there is a growing appreciation that fast breeders are an overelaborate way to make low grade heat: see, e.g., D. Jeambar, "Energie: à l'assault du bastion EDF," *Le Point* (10 January 1977), pp. 44-45. Nonetheless, and despite official denials, Electricité de France is still advertising resistive space-heating installations.

[12] There is no imperative to use the full thermodynamic potential of, say, sunlight for applications not requiring it. Of course it is elegant to use sunlight first to make electricity, then process heat, then space heat; but it may not be necessary or justifiable. Just as failure to capture the full quantity of sunlight available to a given area does not waste anything depletable (except insofar as it may increase collector costs to do the job), so failure to capture the full quality available does not waste anything except a technical opportunity to do something complicated and therefore satisfying to the technician. In other words, for soft technologies, Second Law efficiency is entirely an end-use concept, and should not be applied as an end in itself in designing supply technologies.

[13] This convenience for the user is a headache for the utility, from which fluctuating loads, as many consumers flick their switches, exact costly and elaborate precautions. From a systems point of view the electricity is not nearly as convenient as it is for the user—which is part of the reason that users must pay so much for the transfer of convenience to themselves.

maintain than electric lights, and in some respects are more pleas-ant[14] ; and machinery driven by compressed air seems in no sense less desirable than electrical machinery.[15] Convenience, however one defines it, can also be dearly bought with both money and vulner-ability. Perhaps people are getting to the point where they would rather have the trouble (or, depending on viewpoint, the creative task) of drawing the curtains at night, putting up stormsash, or even adjusting solar plumbing each season than of paying an exorbitant utility bill (on pain of being turned off) or living next to an unwanted reactor.

Badly designed soft technologies can be time consuming, and some commentators suggest that all soft technologies must make excessive demands on one's limited quota of lifetime. But utility systems can be at least as time consuming in terms of the amount of work that must be done to earn the money to pay the utility bill. Further, it is incorrect to argue that all soft technologies must be fiddly and entail onerous personal involvement. To the extent that this is historically true, it is largely because most existing "soft hardware" is the work of enthusiastic amateur tinkerers who enjoy understanding their own systems and increasing their own independence, rather as some peo-ple enjoy camping or messing about in boats. But soft technologies can be at least as "hands off" as the domestic systems that we have and have to fix—today. Such a standard of trouble-free design and construction is indeed assumed in the cost estimates of Chapter Seven. Systems requiring more attention would be much cheaper.

There is no engineering or social reason why soft technologies can-not be exceedingly reliable: more so than hard systems, which have more to go wrong. We do, after all, have some solar heating systems that have worked with zero maintenance for thirty years. The man-agement and maintenance issues are real, and obviously the bad en-gineering that has plagued heat pumps, big turbogenerators, reactors, etc. could in principle occur in simpler systems. (Many episodes of corroded solar collectors come to mind.) But the impact of soft tech-nologies on personal time and care has been much exaggerated.

---

[14] They are efficient, finely controllable, and attractive in color temperature. It is not obvious whether they are more dangerous. Faulty electric wiring is the number two cause (after smoking) of the roughly 12,000 annual U.S. deaths in about three million fires.

[15] Compressed air hand tools are widely used not only in the mining industry (because they are sparkless, rugged, and resistant to wet and dusty environ-ments) but also in, for example, assembly line and machining industries because they are safe, cheap, need little maintenance, are not hurt by being stalled, have good torquing characteristics, and are unlikely to be stolen because most people lack the compressed air supply to use them.

Generally we would need only to return to the standard of engineering and quality control characteristic of many domestic appliances in the 1930s and 1940s. If we cannot do that, there is no chance whatever of making hard technologies work. If we can, we should be able to do it on a small as well as a large scale: the more so because to the person who makes it in the local shop that its small scale permits, the soft-technology device is not a self-contained technical achievement, soulless and without obvious relevance to real people, but will on the contrary be used by people one knows in one's own area.

Is this argument about the marginal—not historical—convenience of electrification academic? Seemingly not. The trends that brought such rapid electrification to supply of the order of a tenth of the primary energy in the industrial world—notably the steep fall in real prices[16] maintained until about 1970—have come to an end, and we need to decide whether there are good reasons nevertheless to divert enormous resources to continuing the task. Major policy decisions, such as whether to proceed with the LMFBR program, are proposed today on the basis that, say, U.S. electricity "demand" will increase fifteenfold, or at least a "conservative" 7.5-fold, by 2020—for what purposes is not made clear. On recent ERDA projections, late-1975 U.S. nuclear capacity is to increase eleven- to twenty-fold by 2000. Von Hippel and Williams,[17] however, have cogently argued that no economic or engineering grounds can justify a nuclear capacity increase above, say, about sixfold (250GW[e]) by 2000, and that ERDA projections showing the contrary are spherically senseless. Indeed, it appears[18] that cogeneration in 2000 could provide[19] electricity conservatively equivalent to over two-thirds of the total 1975 electrical demand, directly displacing over 200 GW(e) of projected U.S. nuclear capacity. The same appears to be true in most other industrial countries.

If existing centralized systems do not now make economic and engineering sense, why were they built? There are at least four rational explanations. First, because objective conditions have changed drastically,[20] and an industry not noted for quick and imaginative responses has been slow to adapt. Second, because centralized energy systems have been built by institutions in no position to ask whether those systems are the best way to perform particular end use func-

[16] See Chapter Six, note 39.
[17] *Ibid.* The same case is made even more strongly in the same authors' March 1977 testimony to the USNRC "GESMO" hearings.
[18] *Ibid.*
[19] See Chapter Two, note 16.
[20] See Chapter Three, note 4.

tions—an omission reinforced by our consistent underpricing of all forms of energy (See Chapters 1.4, 1.5, 3.2) and by utility regulation which automatically increased profits in proportion to capital invested. Third, because at times powerful institutions[21] have deliberately sought to reinforce their power by constricting consumer choice, as in the classic monopoly tactics of the early electric utilities[22] or the fight against public power and (abroad) private wind machines. Fourth, because the long economic shadow cast by large sunk costs has often led us to seek to reinforce past mistakes though subsidies, bailouts, etc., thus further restricting consumer choice,[23] rather than writing off (or gradually retiring through attrition) ill-conceived infrastructure.

Energy decisions are always implemented gradually and incrementally. Major shifts take decades. A chief element of strategy, inherent in the soft path, is thus to avoid incremental commitments of resources to inflexible infrastructure that locks us into particular supply patterns for more decades thereafter (see Chapter 5.5). We are already stuck with gigantic infrastructure that contrains our choices, but we can redirect our effort at the margin, not constrain our choices further. If an average electrical facility lasts forty years, then about half our 1977 stock will not retire until 2012, and stations ordered in 1977 will not retire until about 2027. We must therefore consider carefully how to turn the present stock of big systems to our advantage as a transitional tool, rather than seeking to multiply them in an era when they are no longer what we need more of.

What made sense on the up side of the Hubbert blip, when real costs of electricty (both average and marginal) were steadily falling,[24] may need to be reversed on the down side of the blip, when real costs are rapidly rising with no end in sight. The key problem is at the margin, where we must build new systems that integrate into, or gradually supplant, the systems we already have. Big high technologies probably have their place, and as a former high technologist I would not deny them that place. I merely conclude that it is a limited place and that they have long since saturated and overreached it. The future demands different investments for different needs.

Discussions of this kind are often construed as an attack on elec-

---

[21] Whose "natural monopoly" in legal theory is founded in part on a claimed economic advantage which the above argument challenges.

[22] S. Novick, *Environment 17*:8, 7–13; 32–39 (November 1975), excerpt from *The Electric War: The Fight Over Nuclear Power* (San Francisco: Sierra Club, 1976).

[23] See Chapter Two, note 10.

[24] *Supra* note 20.

tricity and electrification. This is incorrect. The criticisms made here apply to the misuse of electricity—a costly and, as the French would say, noble form of energy—and to the erroneous belief that centralized electrification is an appropriate marginal investment, particularly in industrial countries. Much the same objections apply to central station solar electricity at the margin as to central station fossil or nuclear electricity, for the same economic and social reasons. Some analysts believe, probably on good grounds, that we may well have a large solar-thermal-electric scheme operating in the late 1980s at a cost roughly comparable to that of nuclear power at that time (or, in Central Europe, of oil-fired stations today[25]), with land requirements of the same order as those of equivalent coal or nuclear fuel cycles[26] but able to coexist with some other land uses and to avoid serious scars or impacts.[27] What is not clear, however, is why one should want such a system. It appears that *no* solar electric technology is necessary to U.S. energy supply if one takes full advantage of the diversity of soft technologies already demonstrated.

Interest in solar electricity is unfortunately reinforced by ERDA's arbitrary and inexplicable 1975 decision (ERDA-48) that only electrified technologies—which now receive some two-thirds to three-quarters of all ERDA funding—will be considered capable of making a major long-term contribution. This insistence that 92 percent of U.S. end uses (see Chapter Four) must conform to the supply patterns appropriate to the other 8 percent leads to all sorts of absurd and sterile arguments about which kind of power station to build, not how to meet our end-use needs. Debating whether nuclear or coal-electric or solar-electric stations are cheaper is rather like debating what is the cheapest brand of champagne in town—when what one really wants is a glass of water.

Put differently, the illogic of the ERDA position is this: if we are running out of oil and gas but do not like coal, it is said, we need nuclear power; but if we are not going to have nuclear power, we need other systems that will do what nuclear stations would have done—namely, deliver GW blocks of electricity. But we should instead be seeking systems that will do what we would have done with the oil and gas if we had had them in the first place. It is the function that interests us, not substituting for reactors.

[25] N. Weyss, RR-76-1 (Laxenburg, Austria: IIASA, 1976), pp. 101–170 (and *Intl. Her. Trib.*, 8 March 1977, p. 5, reporting an even more favorable Swiss study by Battelle).

[26] *Ibid.*, pp. 3–4; unpublished manuscript, Dr. Kees Daey Ouwens, Rijksuniversiteit Utrecht, 1976; and somewhat similar, though much less persuasive, results from the Institute for Energy Analysis, Oak Ridge, Tennessee.

[27] See Chapter One, note 15.

Radically decentralized electricity generation could be a very different matter. If, for example, cheap photovoltaic cells are developed—and some analysts believe they have already been developed[28] —they would be extremely useful. Especially if they were cheap relative to direct fuels rather than only to central station electricity, photovoltaics could even be used, without social objection, to increase the range of functions now performed by electricity. (This would probably require, for political and structural reasons, that end uses be changed to low voltage DC and buildings be largely decoupled from the grid, rather than using AC inverters and reversible meters to use decentralized photovoltaics to run the grid; both approaches are technically and economically sound with present technology.) This development may be imminent and should be closely watched, though the analysis in this book nowhere assumes this or any other solar-electric technology. Likewise, decentralized

---

[28] The commercial price usually quoted for photovoltaic arrays in early 1977 is $16/W peak. The money—more than M$100 over the next five years—devoted to lowering this cost appears, however, to be bearing fruit much faster than any one expected, not only through the development of tracking and nontracking concentrators, but of collectors even cheaper than ribbon-grown silicon crystals. Gerald Leach (International Institute for Environment and Development, London) reports that Patscentre International (Melbourn, Herts., U.K.) believe they can produce with present technology a complete CdS photovoltaic collector (using a specially treated 2 μm film), about 4 percent efficient and rather insensitive to angle, for a commercial price of £5.8/m^2. (The prime cost is about half as much; mid-1976 £ $\sim$ $1.70.) At U.K. insolation levels averaging about 1 MW–h m^{-2}y^{-1}, or $\sim$ 114 W/m^2, but with large seasonal variation, this price corresponds to about $0.28/W peak, a highly competitive price. The Shell-financed University of Delaware project under Professor K. Boer is said to be working in a similar CdS price range and planning to pilot produce and market cheap cells with $\sim$ 7–8 percent efficiencies in 1977 or 1978. If neither report turns out to be true, they might well become true in the next year or two because of the unexpectedly rapid progress being made with amorphous and polycrystalline silicon, which has the virtue of being more efficient and nontoxic. The sort of amorphous silicon research reported in the literature (e.g., D.E. Carlson & C.R. Wronski, *Appl. Phys. Lett.* 28:671–73 [1976]) does not capture the excitement of many workers in the field. Likewise a West German group is said to have made coarse polycrystalline Si (grains $\sim$ 1 mm across) of 12.5–13 percent efficiency and to share the widespread belief at the November 1976 Baton Rouge photovoltaics conference that highly competitive photovoltaic cells will be commercially available within a few years, supplying electricity at $\sim$ 2–4 ¢/kW–h(e). (See *Physics Today*, January 1977, pp. 17–19, and *Electronics*, 11 November 1976, pp. 86–99, for a cross-section of Baton Rouge reports.) Even a relatively conservative NAS panel (see Chapter 7, note 22) concluded (pp. 28–29) that "solar cell modules with a cost of a few thousand dollars per average kilowatt (down by a factor of about 50), and having conversion efficiencies approaching 15 percent, could be available by 1980." Such an improvement in conventionally cited photovoltaic costs and performance is in some ways less dramatic than that claimed for future fast breeder or fusion reactors compared with their prototypes.

fuel cells[29] may have a significant role to play, especially in the transitional period. But in essence our energy problem is not a problem of electricity at all, but of liquid vehicular fuels and of clean heat. We cannot hope to address it properly if we continue to devote our resources at the margin to a disproportionately costly and difficult kind of special purpose energy of which we already have nearly twice as much as we can use to advantage.

[29] Fuel cells can in principle be mass produced in simple modules with short lead times, can readily follow 25–100 percent loads, gain dispersion credits (see Chapter 5.2), and can use cleanly a rapidly expanding range of fluid fuels. On present plans (see R. Mauro, Chapter Five, note 19), United Technology Corporation is to demonstrate a 1 MW(e) unit in 1977 and put a 4.8 MW(e) unit on a utility grid in 1979. That module in turn is to form a 26 MW(e) unit of which fifty-six are to be installed on grids starting in 1980, at an initial capital cost of \$250/kW(e) including reformer, cells, and inverter. The cells are to last 40k hours and have a replacement cost of \$50/kW(e). First Law efficiencies are intended to rise from 37 percent in 1980 to 47 percent in 1985 (54 percent counting the $\sim 177°$C waste heat), both with phosphoric acid fuel cells, and to 63 percent (counting the $\sim 538°$C waste heat) in the late 1980s with molten carbonate cells, which are hoped to cost \$200/kW(e) (1975 dollars) in 400 MW(e) + 150 MW(t) units. Mauro also points out many advantages of organic Rankine bottoming cycles, Stirling cycles, and other small- and medium-scale electrical technologies, chiefly for the transitional period. Many such devices, with special reference to solar applications, are described by Office of Technology Assessment, U.S. Congress, "Application of Solar Technology to Today's Energy Needs," 1977.

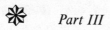 *Part III*

Toward a Durable Peace

Chapter 9

Sociopolitics

9.1. ENERGY AND SOCIAL STRUCTURE

The third main part of this book is concerned with peace: peace between different groups in our societies, different values within ourselves, and different peoples in an increasingly unstable world polity. Yet to seek peace we must first search for tension, and, where we find it, seek to understand its roots.

The dichotomies of social structures and values between soft and hard energy paths define those paths even as they offer us a vivid metaphor for the wider tensions in our lives. It is not the technical or economic but the sociopolitical implications of energy paths that are paramount both in moral importance and in practical impact on political acceptability. The structural tensions between the soft and hard paths are definitive and crucial. Whether an energy system is high or low energy is derivative and less interesting. In energy strategy as in system dynamics, the *structure* of the system matters far more to its behavior than the exact values assumed for its coefficients.

Some commentators, seeking artificially to separate public social issues from private economic issues, seem not to have grasped fully the significance of the hard/soft dichotomy. Alvin Weinberg, for example, has proposed that a strongly nuclear-electric future, modified by "technical fixes" such as clustered underground siting and administration by a technological priesthood socially isolated from the public, should be acceptable to nuclear critics if at the same time the demand for energy is greatly reduced through improvements in end-use efficiency. Let us suppose, though there is no basis for doing so, that the problems of fission are amenable to such "technical fixes" without untoward social side effects. Let us further suppose that a "strong conservation plus strong nuclear" policy is internally consistent, though as a matter of empirical sociology these two constituencies overlap very little: the view that nuclear power is acceptable is normally associated with all the other perceptions of the hard path. Yet even on these assumptions, the argument misses the point: that an essentially nuclear-electric future, even if scaled down, is still a hard energy path in structure and in socio-political implications. The structural effects of pervasive centralized electrification matter far more than whether its enormous extent is doubled or halved. Thus analysts who, like Dr. Weinberg, arbitrarily assume that a nuclear moratorium only means building big fossil-fueled power stations instead of big nuclear power stations[1] are automatically barring themselves from examining any policy questions of real interest.

These questions arise from the high political costs ascribed to the hard energy path by Chapter 2.9. The list is impressive: the hard path, it was argued, demands strongly interventionist central control, bypasses traditional market mechanisms, concentrates political and economic power, encourages urbanization, persistently distorts political structures and social priorities, increases bureaucratization[2] and alienation, compromises professional ethics, is probably inimical to greater distributional equity within and among nations, inequitably divorces costs from benefits, enhances vulnerability and the paramilitarization of civilian life, introduces major economic and social risks, reinforces current trends toward centrifugal politics and the decline of federalism,[3] and nurtures—even requires—elitist technocracy whose exercise erodes the legitimacy of democratic government. Several points of this indictment deserve to be amplified or qualified, and this chapter will do so.[4]

[1] Institute for Energy Analysis, "Economic and Environmental Implications of a U.S. Nuclear Moratorium 1985-2010," draft (Oak Ridge, Tennessee, August 1976).
[2] R.A. Falk, "Rejecting the Faustian Bargian," *The Nation*, 13 March 1976, pp. 301-305.
[3] Some would consider this a good thing. I refer in a relatively value-free way to interregional conflict that does nothing to enhance interregional diversity, and to local versus central conflict that wastes time and effort without producing better decisions or a more enduring political balance. Some people, particularly Marxists, might also consider other items in the above list of political costs to be benefits; that is a legitimate view with which, as a Jeffersonian, I disagree.
[4] See the complementary but strikingly convergent analyses of Lindberg, Hammarlund, and Fritsch (Chapter One, note 12).

In a soft energy path, the technological measure to be achieved can be readily separated from the policy instrument used to encourage it. The former—cogeneration, bioconversion, insulation—is in itself relatively neutral; the latter—taxes, standards, exhortations—may be politically charged. It is only the latter that is likely to irritate us if ill-conceived. But the ends sought are so fine grained, locally tailored, dispersed, and small scale, and the means—the policy tools—can be chosen, according to practical and ideological convenience, from such an enormous array of options, that the choice can fully respect pluralism and voluntarism. Indeed, so diverse are our societies, and hence the local conditions to which soft path innovations must adapt, that a centralized management approach to a soft path simply would not work.

In a hard, coarse-grained energy path, such pluralism would be neither desirable nor possible. The scale and the technical difficulty of the exacting projects are vast. Similarly vast concentrations of social resources must therefore be efficiently mobilized without substantive regard to diverse opinions or circumstances. Only large corporations, encouraged by large government agencies, using large sums of private and public money to employ large numbers of workers on large areas of land, and shielded by the State from the vagaries of the economic and political marketplace, can possibly get the job done. It is not a pluralistic task for householders, small businesses, block associations, or town meetings. It is a monolithic enterprise that demands sweeping, uniform national policies specially devised, with local efforts only an an instrumental and supporting role. The gargantuan organizations involved tend to accrete great power—not only power given them by virtue of their supposed public utility,[5] but further power that they subsume by feeding upon their very size.[6] That size in turn leads to such familiar structural effects

---

[5] Such power has long accrued to energy utilities. Many electric utilities, for example, compete unfairly for capital with other sectors (a form of oblique subsidy) because they enjoy special tax privileges, special bookkeeping conventions, and the ability to sell at a cost rather than a price. That there is a limit to this process is illustrated by recent public statements by Mr. Robert Taylor, Chairman of Ontario Hydro (one of the world's largest utilities), that capital shortage—not projected demand—is *the* dominant determinant of his long-term planning.

[6] Utility size has important antitrust implications. According to L.W. Phillips ("Antitrust in the Electric Power Industry," in A. Phillips, ed., *Promoting Competition in Regulated Markets* [Washington, D.C.: Brookings Institute, 1975]), competitive utility firms produce cheaper power than monopolies only within certain limits of distribution area, and lose that advantage as the area increases. With GW(e) blocks of capacity, the area cannot fail to be large and the operator enjoys a natural monopoly of busbar electricity. It then becomes possible to eliminate competition from, e.g., cogeneration by manipulating rate

as internal alienation[7] and external inequity.[8]

The giant energy facilities essential to the hard path are arcane, remote, unfamiliar, and so overwhelmingly impressive as to be threatening. Huge sums are at stake, experts argue over esoteric technologies and unimaginable risks, national interest steamrollers local doubts. It is not surprising, then, that these facilities breed public distrust and alienation (hence opposition, costly in these escalatory times) by denying public participation not only in the procedural but in the psychological sense.

In contrast, soft energy systems have an obvious relevance to everyday life because they are both physically and conceptually closer to end uses. For example, the New Alchemy Institute's biologically sophisticated Ark project on Prince Edward Island (Canada) has taken hold on the popular imagination all over the province, tapping a hitherto unsuspected reservoir of intelligent interest and initiative. This is not only because the Ark is easy to visit and understand,[9] but also because it is a farm that grows fish and tomatoes—a readily assimilated extension of everyday agrarian life.[10] It makes more sense and seems more relevant to the islanders than a reactor in New Brunswick or a pipeline in Mackenzie, just as Biharis could justifiably suppose that having methane digesters and solar cookers in their own villages would help them more directly than building more reactors in California.

In short, hard technologies are oriented toward abstract economic services for remote and anonymous consumers, and therefore can neither command nor allow personal involvement by people in the

---

structures and refusing to transmit ("wheel") power across the territory controlled. Further consolidation arises from preferential access to capital markets and increased political and regulatory influence. A general legal and policy argument on oligopolistic forces and their relationship to size appears in Mr. Justice Brandeis's celebrated and cogent dissent in *Liggett Co. v. Lee*, 288 U.S. 517 (1933), notably at 548-54 (reviewing the astonishingly recent and stringent legal limits on U.S. corporate size), 557, 565-71, 574 ("Businesses may become as harmful to the community by excessive size, as by monopoly or the commonly recognized restraints of trade"), and 580. See also his earlier book *The Curse of Bigness*.

[7] Early in 1976 *The Times* (London) noted editorially that British companies lose working time through strikes at a rate directly proportional to their size—plausibly a measure of worker alienation.

[8] For an incisive and important discussion of scale and equity, see D.H. Meadows, *Not Man Apart* 6:17, 1 (October 1976), Friends of the Earth (124 Spear St., San Francisco, California 94105).

[9] The Ark also makes manifest our interdependence with the natural world, reintegrating us into it and enhancing our sense of wholeness: a special strength of combined innovation in energy and food systems.

[10] The New Alchemy Institute publishes a journal describing the Ark and many other remarkable projects: NAI, Box 432, Woods Hole, Massachusetts 02543.

community they serve. Soft technologies, on the other hand, use familiar, equitably distributed natural energies to meet perceived human needs directly and comprehensibly, and are thus, in Illich's sense,[11] "convivial" to choose, build, and use.

Of course, there may be Economic People who want as little as possible to do with their own life support systems, and are content to pay their utility bills without a murmur, gobble precooked plastic food as the television exhorts, and eagerly turn every aspect of life into a prepackaged component of GNP. But few such people, free of any subliminal discontent or malaise or alienation, are now observable. Real people generally want to understand their own world and feel responsible for their own destinies, not be mere economic cogs. That is why, for example, even the most materialistic among us complain about impersonal and shoddy "service,"[12] arrogantly paternalistic utilities, incompetent auto mechanics, outrageous bills for simple repairs, and petty bureaucracy. That is why many increasingly cherish the small corner shop, feel more guilty than satisfied at eating hamburgers that are a byproduct of petrochemicals manufacture, and try to persuade the dentist to explain, as one adult to another, exactly what is wrong with that tooth. That is why creative personal activities of all kinds are flourishing—from gardening and canning to weaving and do-it-yourself carpentry (and, arguably, citizen's band radio). People from all walks of life are changing their own fuses, making their own preserves from their own fruit, sewing their own clothes, and insulating their own attics, not only because it pays but because it symbolizes a small triumph of quality over mediocrity and of individualism over the System. The emotions that such involvement releases are powerful, lasting, and contagious. They are a key to understanding the political attractions of soft energy paths.

Soft technologies, then, are inherently, structurally more participatory than hard technologies. Conversely, soft technologies are also less coercive. In a nuclear society, nobody can opt out of nuclear risk. In an electrified society, everyone's lifestyle is shaped by the homogenizing infrastructure and economic incentives of the energy system, and, from the viewpoint of the consumer, diversity becomes

[11] I. Illich, *Tools for Conviviality* (London: Calder and Boyars, Ltd., 1973), and *Energy and Equity* (London: Calder and Boyars, Ltd., 1974). See also D. Livingston, "Intermediate Technology and the Decentralized Society" (Troy, New York: Department of History & Political Science, Rensselaer Polytechnic Institute, 1976).

[12] A recent survey showed 40 percent of consumer purchases resulting in complaints (mostly in vain). The costs of correction, if any, were of course passed through to the consumer via Galbraith's "cost-plus economy," and the only incentive to to better—a powerful one, but slow—would be public withdrawal of legitimacy from the institutions responsible, the last sanction of the people.

a vanishing luxury. Like purchasers of Model T Fords, the consumer can have anything he or she wants so long as it is electrified. But in a soft path, people can choose their own risk-benefit balances and energy systems to match their own degree of caution and involvement. The stakes are smaller, the choices wider, the mistakes more forgiving. Few decisions are irreversible, none compulsory. Preference rules over pattern. People who want to drive big cars or inhabit uninsulated houses will be free to do so—and to pay the social costs. People can choose to live in city centers, remote countryside, or in between, without being told their lifestyle is uneconomic. People can choose to minimize what Robert Socolow calls their "consumer humiliation"—their forced dependence on systems they cannot understand, control, diagnose, repair, or modify—or can continue to depend on traditional utilities.

In a soft path, then, dissent and diversity are not just a futile gesture but a basis for political action and a spur to individual enterprise. Some action by central and local government is of course necessary to get the ball rolling (Chapters 1.5 and 2.4), but then it's mostly downhill: the first push and any later mid-course correction, however laborious they may seem, are slight and brief compared to the uphill struggle otherwise needed to manage the ever-increasing autarchy of a hard path.

One could argue forever whether the structure of the energy system is a cause, effect, or concomitant of social structure.[13] Obviously other influences on social structure, such as information flows, water flows, and land tenure, are important (though often related to the energy system). The skein of causation is thoroughly tangled, and one must not be simplistic: but still a conclusion can and should be drawn. As E.F. Schumacher remarks: "I do not wish to overstate the case; there is nothing absolutely clear-cut in this world, and, no doubt, many tunes can be played on the same piano, but whatever is played, it will be piano music."

It seems undeniable, for example, that energy decisions can and do affect the spatial distribution of jobs, hence of settlements,[14]

[13] R.N. Adams, *Energy and Structure* (Austin: University of Texas Press, 1975); L.K. Caldwell, "Energy and the Structure of Social Institutions," *Human Ecology 4*, 1:31–45 (1976).

[14] Nietzsche remarks that cities that have not decided what they want to be tend to grow to unnatural size. Rome in its glory was slight compared to Athens, for in Rome the energy that should have gone into the flower went instead to make the stem and leaves, and hence was of no consequence. It is a sobering perspective for modern planners who move industrial complexes around on a map like chess pieces on a board. Perhaps the three-to-two ratio of out- to in-migration in US conurbations today relative to the countryside (J. Steinhart, personal communication, January 1977) is trying to tell us something. See Chapter Three, note 8.

hence of political power that can reinforce the pattern. Energy decisions, not only at the point of resource extraction (e.g., the Powder River basin or the Ruhr) but also at the points of secondary conversion, transport, and end use, are unavoidably land use and regional decisions. Just as the railroads once did, and later, perhaps, the superhighways, energy decisions can profoundly and lastingly shape our land use patterns and hence our political patterns. One could at least argue, too, that the extreme capital intensity of new energy systems requires end use to be nearly optimized in an economic sense, so that its nature and patterns conform to the needs of the source of supply rather than the other way around[15]—another influence of energy structure on social structure.

9.2. EQUITY AND POLITICS

Centralized energy systems are inequitable in principle because they separate the energy output from its side effects, allocating them to different people at opposite ends of the transmission lines, pipelines, or rail lines. The export of these side effects from Los Angeles and New York to Navajo country, Appalachia, Wyoming, and the Brooks Range (not to mention Venezuela, the Caribbean, Kuwait, and British Columbia) makes the former more habitable and the latter more resentful. That resentment is finding political expression. As the weakest groups in society, such as the native peoples, come to appear to stronger groups as miners' canaries whose fate foretells their own, sympathy grows for the recipients of the exported side effects.

Throughout the world, central government is trying to promote expansionist energy policies by preempting regulatory authority, and in the process is eliciting a strong state (or provincial) and local response. Washington, Ottawa, Bonn, Paris, London, and Canberra are coming to be viewed locally as the common enemy. Unholy alliances form. Perhaps Montana[16] might mutter to Massachusetts, "We won't oil your beaches if you won't strip our coal." As Congress—made of state people with no federal constituency—increasingly molds interregional conflict into a common states' rights front, decisions gravitate by default to the lower political levels at which consensus is still possible. At those levels, further insults to local

[15] I favor peak load pricing, and possibly such measures as ripple signal load control for some applications; but one can hardly pretend that load management techniques (see Chapter Four, note 21) are not intended to alter the patterns of use and in some sense—one of which undersized solar water heaters are often accused—to alter the consumer's lifestyle.

[16] B. Christiansen & T.H. Clack, Jr., *Science 194*:578-84 (1976).

autonomy by remote utilities, oil companies, banks, and federal agencies are intolerable.

Thus people in Washington (and their counterparts overseas) sit drawing reactors and coalplexes on maps, but the exercise has increasingly an air of unreality because it is overtaken by political events at the grass roots. The greater the federal preemption (as in offshore oil leasing), the greater the homeostatic state response. The more the federal authorities treat centrifugal politics as a public relations problem, and the more they take the authoritarian view (as in West Germany) that local objections must be stifled by national imperatives, the more likely it becomes that they will not only fail to get their facilities built, but will also in the process destroy their own legitimacy. To some extent this has already occurred. It seems likely, for example, that U.S. states will soon gain a veto power, at least, over nuclear facilities in their jurisdiction, as current federal legislation proposes and as some states have already enacted—a process being echoed throughout Europe. On this issue as on others, the traditional linear right-left political spectrum seems to join to form a circle as differently grounded distastes for big government merge across gaps of rhetoric. The resurgence of individual, decentralist citizen effort in politics, as in private life and career, seems an important political universal in most industrial nations today.[17]

Big Brother does not like losing his grip. In 1975, for example, some U.S. officials were speculating that they might have to seek central regulation of domestic solar technologies, lest mass defections from utility grids damage utility cash flows and the state and municipal budgets dependent on utility tax revenues. Since utilities are already perceived as having too much power and utility regulators too little sensitivity, a surer recipe for grassroots revolt would be hard to imagine. Perceptions of the value of dependence on utilities may be shifting rapidly as the enterprise reaches such a size that it starts to intrude on life in many traditionally "safe" areas, as in Ontario and the Orkneys; or as its vulnerability becomes painfully manifest, as in England; or as general political consciousness rises in step with utility bills.

The disillusionment and resentment one can see in many industrial countries is akin, perhaps, to that of a citizen of a poor country who

---

[17] D.E. Lilienthal, Sr., "America: The Greatest Underdeveloped Country" (Address to a public meeting sponsored by the American Institute of Planners and the American Society of Planning Officials, Sheraton Park Hotel, Washington, D.C., 22 March 1976). The individualistic renascence Lilienthal welcomes could presumably be encouraged, as in the land grant days, by an Energy Extension Service, and in some sense is not far from the concepts of Steinhart (see Chapter Three, note 8).

is realizing that an energy technology predicted to bring self-reliance, pride, and the development of the villages has actually brought dependence, a cargo-cult mentality, and the enrichment of urban elites. If this is right, then the recent shift of institutional and individual investment away from utilities reflects not only concern with debt structure and interest coverage but also, more fundamentally, with reduced attractiveness or expectations—in an intuitive sense of which investors are keenly aware—because of gradual withdrawal of legitimacy by a fickle public that has already done the same to oil majors. The grounds of this shift among a previously tolerant, even supportive public are structural, arise essentially from justified suspicion of centrism, and would not be reversed by nationalization or rechartering that ignored scale. As Nowotny notes in a preliminary suggestion of one of many causes,

> ... opposition against nuclear power in its social structures roots is *opposition against those who will benefit from further economic and political concentration and centralization.* It is directed against "big" industry, seen in collusion with "big" government and "big" science. It is the opposition coming from those who feel *powerless and small* in the face of these developments.[18]

Anyone living in England as I do must be acutely aware of the vulnerability of an energy system that permeates the end-use structure, distributing a form of energy that cannot readily be stored in bulk. Its supply relies on hundreds of large and precise machines rotating in exact synchrony across a country or a continent, strung together by a vast aerial web of frail arteries that can be severed with a rifle. Disruption can be instantaneous and pronounced: so much so that English users have gone to very costly lengths to protect themselves, reportedly installing more private than public capacity in some recent years despite substantial obstacles.

The vulnerability of the grid is enhanced not only by the guerrilla activity[19] endemic in Britain but also by the capital intensity of the grid. In the U.S., a power station requires roughly a quarter of a million dollars' investment per direct job. Correspondingly low labor intensity and technical specialization (high skill intensity) mean that political power resides in fewer and fewer workers whose highly

---

[18] H. Nowotny, "Social Aspects of the Nuclear Power Controversy," RM-76-33 (Laxenburg, Austria: IIASA, April 1976), p. 25. (Emphasis in original.)

[19] B.M. Jenkins ("High Technology Terrorism and Surrogate War: The Impact of New Technology on Low-Level Violence," P-5339 [Santa Monica, California: RAND Corp., January 1975]) suggests this may be the future of warfare because conventional war is too expensive.

156 Toward a Durable Peace

developed skills may be matched by coherent purpose and organization. Most power engineers appear to have an admirable sense of dedication to the public good. In Britain, however, a leader of the power engineers recently remarked—correctly—that "the miners brought the country to its knees in eight weeks; we could do it in eight minutes." His union, before starting a recent round of wage negotiations, even proposed a twenty-four-hour initial strike as a token of sincerity. And as other English experience has shown, for example with Windscale workers and National Health Service doctors, the attitudes of organized groups of key people can change quickly and decisively. In this sphere as in the military sphere, the nature of a technology has fundamentally altered the power balance between large and small groups,[20] making it less tolerable for potentially hostile or abnormal people to enjoy traditional civil liberties,[21] and generally increasing the level of suspicion and intolerance in society. "Power to the people" now makes us ask: "Power to *which* people?"[22]

Such political costs, prominent in the list of nuclear issues, correspond to analogous costs of possessing military nuclear weapons: indeed, the latter costs may have helped to sensitize us subliminally to the former. For example, nuclear weapons demand an elite, self-perpetuating cadre that will meticulously and perpetually guard and prepare the weapons, isolated alike from social unrest and from proximate (effective) political accountability, rather like Dr. Weinberg's concept of a "technological priesthood" for guarding long-lived nuclear wastes.[23] Nuclear weapons demand a fast-reacting central authority that can make decisive and timely military responses without the niceties of balanced democratic consultation, just as nuclear power tends to weaken political responsibility through centrism and heavy technological content. Further, the power of nuclear weapons requires that willingness to use them be periodically demonstrated obliquely—by fighting small nonnuclear wars as theater,[24] not to attain classical military objectives so much as to present an image of united national will. This image may require public dissent at home to be suppressed, or prevented by deception,[25] and thus contributes both to loss of public candor and trust and to the evolu-

---

[20] *Ibid.*

[21] M. Flood, *Bull. Atom. Scient. 32*:8, 29–36 (October 1976); Chapter Two, note 34.

[22] *Ibid.*

[23] See Chapter One, note 18. Dr. Weinberg's review and a response appear in *Energy Policy* (U.K.) *4*:363–6 (December 1976).

[24] J. Schell, "The Time of Illusion: VI. Credibility," *The New Yorker*, 7 July 1975.

[25] *Ibid.*

tion of an executive siege mentality—events with an obvious parallel in the civilian sector. In several fundamental ways, then, the military imperatives of nuclear weapons conflict with the political imperatives of a free society. Latent public unease with this conflict may be an important indirect element in the political climate[26] in which opposition to reactor siting proposals can flourish.

9.3. INSTITUTIONAL PROBLEMS

Hard technologies bring a little appreciated but important array of pressures to bear on individual high technologists. Intense social pressures within committed institutions tend to discourage deviance from shared values and beliefs, reinforcing them with apparent consensus (deserved or not) even at the expense of personal ethics. It is well known that people in groups, even with high intentions, do things they would never dream of doing individually; and our professions have not yet devised ways to shield the delicate spirit of inquiry and dissent from subtle (or even not so subtle) peer pressures. It is common, for example, in nuclear safety programs to find excellent technologists whose public and private views perforce must differ, so corroding both personal and technical quality.

Shifting one's career to a field that one can have good dreams about may not be an attractive option: so long as federal R&D interest is perceived to be mainly in hard technologies, good technologists will be reluctant to shift to soft ones, thus limiting the long-term ability of those sectors to absorb funding effectively and further reducing their attractions. This is a practical demonstration of the exclusivity of energy paths (see Chapter 2.11): no more eloquent example can be given than the way that the dominance of fission (especially breeder) projects in energy R&D programs throughout the world in the past few decades has left us today with a restricted range of options and an inflexible set of technical institutions.

The big organizations needed to develop big technologies often suffer from fuzzing of responsibility. So many people contribute to a project that nobody is really responsible,[27] and responsibility in the traditional engineering sense may slip through the cracks. But more destructive to technical quality is the tendency of such big organizations, after the exciting surge of initial pioneering, to become routine and deadly serious. For that reason alone—to say nothing of public controversy—fission is now no fun any more; therefore

[26] J.C. Mark, "Global Consequences of Nuclear Weaponry," *Ann. Rev. Nucl. Sci.* (in press).
[27] A.J. Ackerman, *Trans. IEEE AES-8:*5, 576 (1972).

it will not be done well; therefore it has failed.[28] A subtle but important disadvantage of big technologies (as Freeman Dyson points out) is that they are so big that one cannot play with them, so an essential breath of both fun and creativity is lost.

In contrast, soft technologies bring out the ingenious tinkerer, the spirit that made the sophisticated (but understandable) gadgets that fill any good farm museum. The challenge of simplicity, the art of artlessness, have full scope. Human scale relative to the technologist lets ingenuity—the source of paper solar collectors,[29] heat cables,[30] and exterior retrofit insulation[31] —expand far beyond the narrowly focused limits attainable in hard technologies.

A further institutional problem from which soft technologies are nearly exempt is control of potentially hazardous developments. Soft technologies are by definition nonviolent. Hard technologies are often Siamese twins of weapons (see Chapter Eleven, or consider solar satellite schemes or laser fusion[32]). Worse, subjecting civilian or quasi-civilian hard technologies to the necessary political control may require, for informed discussion, the release of information that endangers the public. For example, to provide a basis for intelligent decisionmaking about LMFBR disassembly accidents, one must at the same time reveal a good deal about how to design atomic bombs. In order not to be faced with the impossible choice between endangering the public by releasing information useful to the malicious and, on the other hand, ceding decisionmaking on a crucial matter to an elite[33] and so sacrificing democratic principles, one may

[28] Likewise, perhaps really well-engineered nuclear facilities would be so boring to operate that talented people will seek other careers. Some observers see this as a major long-term problem at facilities like Barnwell (which is not to say it is universally considered well-engineered or will ever operate).

[29] Mentioned in Chapter 7.4; Dr. Plunkett is Managing Director, Montana Energy & MHD R&D Institute, Inc., Box 3809, Butte, Montana.

[30] A.E. Haseler tells me (1976) that flexible, very well-insulated heat mains for distributing hot water to (and returning cooled water from) low density residential areas can now be delivered on cable drums for rapid installation at very low cost, so making district heating economically attractive even for dispersed suburbs. See also *New Scient.* 73:522 (3 March 1977).

[31] I. Höglund & B. Johnsson (Royal Institute of Technology, Stockholm), in their International CIB Symposium paper (see Chapter Two, note 15), describe a method they have developed for insulating existing apartment buildings on the outside, so as not to disturb the tenants or make the rooms smaller.

[32] See, e.g., W.D. Metz, *Science* 194:166 (1976). A good argument can be made that laser fusion research should be abandoned forthwith because of the sensitive knowledge it proliferates.

[33] See Chapter Two, note 35; A.B. Lovins, comments in S.M. Barrager et al., "The Economic and Social Costs of Coal and Nuclear Electric Generation: A Framework for Assessment and Illustrative Calculations for the Coal and Nuclear Fuel Cycles," Stanford Research Inst. MSU-4133 (Washington, D.C.: USGPO, 1976); Chapter Three, note 35; B. Kennard, "We Are All Scientists Now: The

have to head off such technologies as soon as such a choice appears on the horizon. To do this effectively, one may be forced to try to suppress lines of research that are still in the realm more of science than of technology. This in turn raises unpalatable issues of liberty versus license in academic freedom: a controversy now raging in genetic research, which is in some ways closely analogous to the state of nuclear science in the 1940s. Such encroachment is less likely to be considered or needed at all if we channel our efforts into less exploitative, more adaptive, gentler technologies[34].

New Case for Public Participation'' (room 300, 1785 Mass. Ave. NW, Washington, D.C. 20036), 1976.

[34] A simple example from E. Robertson (Winnipeg): There are three ways to make limestone into a structural material. We can cut it into blocks, which is not very interesting. Or we can bake it at thousands of degrees into Portland cement, which is inelegant. Or we can feed chips of it to a chicken. Twelve hours later it emerges as eggshell, several times as strong as the best Portland cement. Evidently the chicken knows something we don't about ambient temperature technology.

Chapter 10

Values

10.1. MEANS AND ENDS

Chapters One, Two, and Nine have sketched a few examples of the labyrinthine complexity with which energy
questions infiltrate into every aspect of our lives. The policy questions that the soft path makes so prominent raise also some more
basic structural and philosophical questions that we can neither
answer definitively nor ignore: questions of the personal values
within which, and the social ends for which, energy is sought. This
chapter will amplify Chapter 2.10's sketch of a few of these questions—economic growth and its diseconomies, distributional equity,
and transindustrial values.

A cursory glance at a graph of U.S. per capita primary energy
(let alone end use energy) since, say, 1850[1] —and likewise in many
other countries—dispels any notion of a significant correlation with
social welfare on a time scale of decades. The graph remains flat for
several long periods when welfare by almost any measure improved
significantly, for example, in 1850-1900 and (on average) 1920-
1950, then rises steeply during the 1960s out of all proportion to
real improvements. In the future, the gap between primary energy
per capita and function performed (not the same as welfare) per
capita would widen even more rapidly, as sketched in Figure 2-3.
Nonetheless the existence of a close correlation, even an ironclad

[1] E.g., J.C. Fisher's graph, reproduced in IIASA RR-76-1 (Laxenburg, Austria:
IIASA, 1976), p. 205.

.ty, is an article of faith to many who measure our success
e quantity of goods and services consumed, rather than by how
ve achieve human satisfaction, joy, and inward growth with a
minimum of consumption—the concept central to Schumacher's
classic essay[2] on Buddhist economics.

The former view is consistent with the industrial era paradigm in
which, as Mumford caustically observes:

> All but one of [the seven deadly] . . . sins, sloth, was transformed into a
> positive virtue. Greed, avarice, envy, gluttony, luxury, and pride were the
> driving forces of the new economy. . . . Goals and ends capable of working
> an inner transformation were obsolete: mechanical expansion itself had
> become the supreme goal.[3]

As a result of this process, as Paul Diesing notes:

> One cultural element after another has been absorbed into the ever-
> widening economy, subjected to the test of economic rationality, ration-
> alized, and turned into a commodity or factor of production. So pervasive
> has this process been that it now seems that anything can be thought of
> as a commodity and its value measured by a price . . . time, land, capital,
> labor; also personality itself, . . . art objects, ideas, experiences, enjoy-
> ment itself, and even social relations. As these become commodities they
> are all subject to a process of moral neutralization.[4]

The very language that weaves our thoughts about the world is
thus conditioned by an economic paradigm that treats means as
ends[5] and places goods above people.[6] But the cult of material
acquisitiveness underlying our assiduously promoted image of the
good life has its flaws, and we are coming to see them more clearly.

For example, we concentrate our resources not on microefficiency
or resilience but on managing our growth and resolving its numerous
conflicts and inequities. So we weave a web of bigness and of incomprehensible, unmanageable complexity. But are we sure that our
transaction costs do not exceed our productivity? Is it worthwhile,

---

[2] See Chapter One, note 14.

[3] L. Mumford, *The Transformations of Man* (New York: Harper, 1956).

[4] P. Diesing, *Reason in Society* (Urbana: University of Illinois Press, 1962).
This and note 3, *supra*, are taken from Chapter One, note 15.

[5] See Chapter Three, note 2; C. Cooper, ed., *Growth in America* (Washington, D.C.: Woodrow Wilson Center/Westport, Connecticut: Greenwood Press,
1976); T.V. Long III & L. Schipper, "Resource and Energy Substitution," in
US Economic Growth from 1975-1985: Prospects, Problems, and Patterns
(Washington, D.C.: Joint Economic Committee, U.S. Congress, 1977); H.E.
Daly, "Entropy, Growth, and the Political Economy of Scarcity" (seminar
paper, Resources for the Future, Washington, D.C., 1976).

[6] *Supra* note 2.

or possible, to go on like this? Miles states thus his thesis that it is not possible:

(1) The more energy a society uses, the more interdependent it tends to become, both within itself and in relation to other societies.
(2) The more interdependent a society becomes, the more complex it becomes, and the more man-designed and man-controlled its economic, ecologic, and political systems and subsystems become.
(3) The more complex and interdependent the systems and subsystems, the more vulnerable they become to design failures, since:
    (a) No human designers, and this applies especially to the politicians who are responsible for designing the largest human systems, can know or comprehend all the factors that need to be taken into account, and their interrelation, sufficiently to make the current set of systems work well. If complexity and interdependence increase further, the problems will be further compounded.[7]
    (b) Those responsible for selecting the designers—the voting public in a society like the United States—are even less informed about the intricacies of the systems than the politicians who represent them. They cannot judge, therefore, which programs or social designers (politicians) to support, and in consequence they are highly likely to vote for the representatives who promise to support programs that benefit them directly and immediately—a fatal flaw in designing workable complex systems for interrelating enormous numbers of human beings with each other and their environments.
(4) The United States is probably nearing the point (it could even be beyond it) where the complexity of the systems of interdependence exceeds the human capacity to manage them, causing system breakdowns to occur as fast as or, faster than, any combination of problem-solvers can overcome them.
(5) World systems of interdependence are more remote, inefficient, and precarious than national systems, and may have exceeded their sustainable level of complexity.*** More and more nations are looking for ways of becoming more independent, rather than more interdependent, even if to do so they will have to be satisfied with a lower standard of living. Those social analysts who have asserted that we will inexorably continue to move upward toward higher levels of technology, higher levels of energy consumption, greater volume of international exchange of raw materials, goods, information, culture, and tourists may be in for a surprise. Humanly designed and operated systems have upper limits of complexity, and when they reach those limits of complexity, they simply break down.[8]

[7] See Chapter One, note 16.
[8] R.E. Miles, Jr., *Awakening from the American Dream: The Social and Political Limits to Growth*, summarized in the *Washington Star*, 9 May 1976. An interesting related study of societal vulnerabilities and interdependencies is being undertaken by Harrison Brown et al., for 1977 publication (New York:

s argument is important on a smaller scale than whole
is the thesis of Kenneth Boulding:

> There is a great deal of evidence that almost all organizational structures
> tend to produce false images in the decision-maker, and that the larger
> and more authoritarian the organization [to attempt to cope with its
> complexities], the better the chance that its top decision-makers will
> be operating in purely imaginary worlds. This perhaps is the most funda-
> mental reason for supposing that there are ultimately diminishing returns
> to scale [and to growth itself].[9]

Here is another conundrum. Assuming that people seek to escape
from repugnant work into blissful leisure, we strove mightily to
mechanize, automate, and fragment work. But while that effort
at first relieved mindless drudgery, it came increasingly to deprive
people of meaningful roles, even of jobs themselves. As the crafts-
person working creatively with tools was displaced by the machine
demanding a stultified operator, people, especially those whom age
or sex or ability ill-suited for economic roles, were deprived of a
share in a visible and widely shared public purpose. We systematically
substituted money and energy for people, calling this "improving
productivity"—by which we meant labor productivity—and then
reinforced this mistake by concluding that we should use still more
labor-replacing high technologies to fuel the economic growth
needed to employ the people disemployed by that very process. We
defined work as obtaining a commodity (a job) from a vendor (an
employer)—so that work itself became a commodity produced, no
less, by a process of production, like a brick or a car. We substituted
earning for an older ethic of serving and caring as the only legitimate
motivation for work. Thus alienation in place of fulfillment, inner
poverty alongside outward affluence, a pathologically restless and
rootless mobility became the symptoms of a morbid social condition
that corroded humane values.

Our brilliant success in achieving economic goals has thwarted
the human goals of much of our population by declaring their lives
to lack value. Might not happy people (in D.T. Suzuki's phrase)
instead enjoy life as it is lived rather than trying to turn it into a
means of accomplishing something else? Our sense of the whole, in
personality and family as in community and country, has been

Norton). See also K.H. Humes et al., "The Future Environment: U.S. and World
Trends," NASA (contracts NAS5-20732 and -34), 15 July 1975.
[9]K. Boulding, *Am. Econ. Rev.*, May 1966, p. 8. For a related thesis (loss of
resolution with increased scope) see J.G.U. Adams, *Envt. and Planning* (U.K.)
4:381–394 (1972).

sacrificed to specialization and mobility, the imperatives of efficient production; but is that price worth paying? Life near enough to the soil to understand its rhythms, balances, and tensions has nurtured every previous culture,[10] and for all its hardships in the past—most of them now avoidable—it has a cultural value; are we sure that by rejecting that awareness of the life process and turning ourselves into mere cogs of abiotic production we are not losing something essential to the human psyche and to our own mythic coherence?

Again, if we work to buy a car without which we cannot get to work, is the *net* benefit so very substantial? Is not this circularity a measure less of our wealth than of our failure to create fulfillment with a sensible economy of effort and time? Ivan Illich[11] cites a number that, whether it is quite correct or not, still conveys an important idea: he says that the average American man drives about 7500 miles a year in his car, but that the total time it takes him to do this and to earn the money to finance it is about 1500 hours, which works out to five miles an hour, and we can walk nearly that fast.

Again, three billion people offer living refutation of a theory that assured us that our growing wealth would automatically enrich the poor without our having to redistribute anything.[12] Spurred by the profoundly important development of New Economic Order concepts, affluent societies are waking up: in the famous Harris poll of 1975, for example, even before many of those concepts became widely known, a solid 61 percent (to 27 percent) thought it is "morally wrong" for Americans to consume such a disproportionate share of the world's resources, and 50 percent (to 13 percent) worried that a continuation of such behavior will "turn the rest of the world against us." The theory persists even on a domestic scale: appeals for more electricity with which to help the poor still come from utilities that have long charged the poor several times as much as the rich (or profligate) for the same unit of electricity.[13] Yet we can now calculate that raising, say, the poorest three-fifths of Americans to the per capita primary energy level of the upper middle class

[10] Agrarian cultures are not too theoretical and tend to have very rational economics. For an example, see "The Return of the Draft Horse," *Organ. Gard. & Farm.*, February 1976, pp. 156–64, and "Times Diary," *Times* (London), 31 March 1977.

[11] I. Illich, *Energy and Equity* (London: Calder and Boyars, Ltd., 1974).

[12] This is also known, with apologies to Marie Antoinette, as the "Let them eat croissance" theory.

[13] Systems of flattened or lifeline utility rates that are desirable for equity may occasionally depart from narrow economic rationality. That is all right: there is nothing sacred about economic rationality. In general, however, energy should be properly priced, and its price is not a suitable instrument of distributional equity. If poor people cannot afford properly priced energy, they should be made less poor in ways that do not entail giving cheap energy to rich people.

not require massive energy growth, but rather a level[14] well
any current official projections, even assuming no efficiency
improvements of any kind. Even more significantly, North Americans can now see that reducing themselves to the "primitive"
energy level of most Europeans—let alone the presumably Neolithic
New Zealanders, at about a fourth the North American level—need
not reduce material comforts at all and could markedly improve the
quality of life.[15] And the argument does not even stop there: most
houses in New Zealand today, for example, are essentially uninsulated, public transport is deteriorating, and energy efficiency is
rather low.

Some consider this line of thought truly subversive of the established order, and particularly of the established economic church.
As Galbraith points out:

> Nothing would be more discomfiting for the economic discipline than
> were men to establish goals for themselves and on reaching them say,
> "I've got what I need. That is all for this week." Not by accident is such
> behavior thought irresponsible and feckless. It would mean that increased
> output would no longer have high social urgency. Enough would be
> enough. The achievement of the society could then no longer be measured
> by the annual increase in Gross National Product. And if increased produc-
> tion ceased to be of prime importance, the needs of the industrial system
> would no longer be accorded automatic priority. The required readjust-
> ment in social attitudes would be appalling.[16]

Proponents of endless energy growth are fond of saying that we
need the energy to accomplish our goals. What goals? Goals imply
that they are bounded, are finite, can be attained; but a goal of
"more" has no end. Whose goals? How much is enough? Enough
for whom? If we cannot say how much is enough, do we need

[14] An estimate by the Union of Concerned Scientists is 117 quads in 2000,
assuming Series E population growth—with a fertility rate higher than today's—
to 265 million by then and no energy conservation of any kind.

[15] See Chapter Two, notes 11–15; and, for a representative sampling, R.
Doctor et al., *California's Electricity Quandary: III. Slowing the Growth Rate*,
R-1116-NSF/CSA (Santa Monica: RAND Corp., 1972); and W. Ahern et al.,
Energy Alternatives for California: Paths to the Future, R-1793-CSA/RF (Santa
Monica: RAND Corp., 1975); A.C. Sjoerdsma & J.A. Over, eds., *Energy Con-
servation: Ways and Means*, publication 19 (The Hague: Stichting Toekomst-
beeld der Techniek, 1974); D.A. Pilati, *Energy 1*:233–39 (1976); Dubin-Bloome's
June 1976 study of public utility rates and energy conservation for the New
Jersey Public Advocate's office; Skidmore, Owings, and Merrill, "Bonneville
Power Administration Electric Energy Conservation Study" (July 1976); *News-
week*, 14 February 1977, p. 24.

[16] J.K. Galbraith, *The New Industrial State* (Boston: Houghton Mifflin,
1967), p. 319.

more? Who has enough already? Do the Germans think they have enough? Do they think the Americans have enough? Do the Indians think the Germans have enough? Who thinks the New Zealanders have enough? Which New Zealanders? How far does tomorrow's growth steal from our children? What would they say about that?

One reason these disturbingly simple questions need asking now is that the factual basis of the world in which they long seemed irrelevant has changed profoundly. The economic growth that served the industrial countries well, according to their lights, took place in an era when they bought raw materials at competitively depressed prices and sold manufactures at monopoly rents. Now, as Jay Forrester has pointed out, they must instead buy raw materials at monopoly rents and sell manufactures at competitively depressed prices. They cannot get out of that hole—buying dear and selling cheap—by expanding their international trade; quite the contrary. And so the whole economic paradigm of growth and free trade is no longer useful to those who have profited most from it.

Still another fatal flaw of the growth and free trade model is that it does not account for regional disparities. Even the industrial countries have chronically poor places like Newfoundland and Calabria. According to the classical model, such places must have a competitive advantage in something on a world scale; or their people must be happy to move to Toronto or Milano; or Torontans and Milanesi must be willing to redistribute a substantial share of their income to Newfoundlanders and Calabresi forever. Recent political experience gives us no grounds for supposing that *any* of these three conditions can be satisfied.

We have also lately discovered that the compelling logic of national balance of payments, in a world where money does not recycle freely and equitably, has an important analogy on a subnational scale provincial, town, even household. The provincial government of, say, Prince Edward Island may well find that it is advantageous to pay more for a product made locally than to import it from Ontario—because buying locally means keeping the money *chez nous*. There is an enormous difference between an internal transaction, which only transfers money within one's own region, and an external transaction that requires one to support middlemen and transaction costs in Ontario. The external transaction also requires one to tie oneself more firmly into Ontario's commercial web so as to earn the equivalent of foreign exchange with which to buy things from Ontario. Yet this difference between internal and external transactions, so obvious to people like the Premier of Prince Edward Island, is not at all obvious to most economic theorists. It is another example of the inappropriateness of most modern economic models to the new circumstances with which we must live from now on.

10.2. TRANSFORMATIONS

As we learn to question the ability of present policy to serve both public and private ends, the legitimacy of those ends themselves comes up for review. Our know-how has far outstripped our know-why; and as we seek to redress the balance, old political concepts begin to reassert themselves. Grassroots democracy acquires[17] a more concrete meaning. Jefferson and Mao gain a curious affinity. Control of property and land—a cornerstone of free enterprise democracy—comes to embrace control of the energies essential to life, liberty, and the pursuit of happiness, for to control those energies we must now control the land they lie under or fall upon. In the process we may start to approach Aldo Leopold's land ethic,[18] or even the native American/Canadian/Australian concept[19] that absolute, monopolistic ownership of land, at first incomprehensible, is in a sense blasphemous.

In our dynamic, pluralistic culture, our decisions will be made at the points of tension between competing value systems.[20] Some of these tensions have long been with us but are now taking on a new

---

[17] See Chapter Nine, note 17; M. Lönnroth, "Swedish energy policy—technology in the political process" (Stockholm: Future Studies Group, 26 October 1976), esp. pp. 9, 49. Today's problems of governance were clearly foreseen in the writings (particularly the letters) of Jefferson, who once remarked that the system of government that he and his colleagues had devised is ideal for a rural agrarian population, but that if Americans ever crowded into cities as the Europeans do, the former would become as corrupt as the latter.

[18] A. Leopold, *A Sand County Almanac* (Oxford: Oxford University Press, 1949, paperback 1968).

[19] This is hardly a new theme:

Moreover the profit of the earth is for all: the king himself is served by the field. He that loveth silver shall not be satisfied with silver; nor he that loveth abundance with increase: this is also vanity. When goods increase, they are increased that eat them: and what good is there to the owners thereof, saving the beholding of them with their eyes? The sleep of a laboring man is sweet, whether he eat little or much: but the abundance of the rich will not suffer him to sleep.

—Ecclesiastes 5:9–12.

[20] See, e.g., H. Kahn, J. Brown, S. Brand, and A. Lovins, discussion, *Co-Evolution Quarterly,* spring 1977. The following portion of text draws on the author's draft contributions to *The Unfinished Agenda*, Report of the Environmental Agenda Task Force, G.O. Barney, ed. (New York: Rockefeller Brothers Fund/Thomas Y. Crowell, 1977).

significance. For example, for many historical reasons, including the forces that selected those who emigrated to the U.S., it is a basic tenet of American social philosophy that economic and social activity are primarily to serve private and individual, not public and communitarian, ends. Public ends are indeed conceived not as abstract moral universals, such as making a good or a just society, but rather as a pragmatic summation of diverse private ends. This value structure, this very individualism that has enabled Americans to accomplish so much materially, has nurtured the primacy of economic and technological values over spiritual and humanistic ones, and has long kept people from coming to grips with their own tragedies of the commons.[21]

As "economic growth and technical achievement, the greatest triumphs of our epoch of history, [show] . . . themselves to be inadequate sources for collective contentment and hope" (Robert Heilbroner), we are starting to appreciate that we have no monopoly on wisdom. If we can suspend our cultural arrogance, we may learn much from other peoples with different values and longer experience. We may, for example, be surprised and much moved to hear the chiefs of the Micronesian island of Maap, threatened by a Japanese hotel complex, petition us thus:

Whereas we love our land and the ways in which we live together there in peace, and yet live humbly and still cherish them above all other ways, and are not discontent to be the children of our fathers, it has become apparent to us that we have been persuaded to subscribe to processes that will quickly extinguish all that we hold most dear.*** We, pilungs and langanpagels, elders and elected officials, Chiefs in Council of and on behalf of all the people of the Eighteen Villages and Fiefs of the Island of Maap . . . declare our love of this place and of the ways passed down to us by the generations. We have inherited from our fathers a land that is lovely and provides for us the fruits of the earth and of the sea. We are few in numbers but have a brave history and are strong in our resolve to preserve these things that are sweet to us and freely to determine the affairs of our island with respect to custom and deference to the law. . . .*** We . . . now urgently and passionately unite . . . to ask the help of . . . [others] in ridding us of this invasion and freeing us, that we do not become servants in our own land, to choose for ourselves the paths that will be good for the people of all the villages of Maap.*** [W]e meet today under the shadow of . . . change and innovation and . . . we know our people to be roused against these things, as they today do forcefully convey through a petition. . . . [W]e are therefore all the more solemnly moved to affirm our united will in the face of the unfamiliar contingencies

[21] G. Hardin, *Science 162*:1243-8 (1968).

this age and the ages to come, so that our home may not be vulnerable
the casual invasions of those who do not know our hearts or the dis-
ɪoyal speculations of those who do.[22]

Such language, so reminiscent of America's national beginnings,
must remind us of the power of timely ideas—republican government in monarchical Europe, or Christianity in Rome—to transform
societies with extraordinary speed. The thesis that we stand today on
the verge of such a transformation must command attention.[23]
Social change today occurs everywhere at dizzying speed. And at
this time of choice in the midst of a crisis of ideals and actions, we
are developing lines of policy that converge with many of the most
important forms of social change while cutting across traditional
lines of political conflict (see Chapter 1.5). Perhaps we are approaching a new vision, a new synthesis. As we start to see, in Alwyn
Rees's phrase, that when we have come to the edge of an abyss, the
only progressive move we can make is to step backward, we begin
to realize that we can instead turn around and then step forward,
and that the turning around—the transition to a future unlike anything we have ever known—will be supremely interesting, an un-
precedented central project for our species.

Faust, having made a bad bargain by not reading the fine print
and so brought disaster on the innocent bystanders (Gretchen's
family), was eventually redeemed and accepted in heaven because
he changed his career, redevoting his talents to bringing soft technologies to the villagers. We need, like Faust, to refashion hubris
into humility; to learn and accept our own limits as a fragile and
tenuous experiment in an unhospitable universe; and to grow content to live as people, not as gods. Our choice of "the road less
travelled by" can truly make all the difference. But if we wish to
have the chance to tell of our choice, "somewhere ages and ages
hence," then we must choose soon, and choose wisely, for all the
ages.

[22] R. Wenkam, Introduction to K. Brower, *Micronesia: Island Wilderness*
(San Francisco: Friends of the Earth, 1976). (The petition was translated,
apparently without embellishment, by American lawyers.) Similar conclusions
can be drawn from many other native cultures: e.g., FOE's earlier book on the
Brooks Range of Alaska, *Earth and the Great Weather*, or the writings of Laurens
van der Post on the Bushman culture.
[23] See Chapter Two, note 36.

Chapter 11

Rebottling the Nuclear Genie

A heavy burden of choice lies on our generation. For the first time in the human experiment, it lies in our power, in our age, to end that experiment for future ages. We now cower under the threat that, through a moment of error, terror, or madness, thousands of small stars will be kindled here and there on the surface of our planet. A few incandescent hours—followed by withering, lingering clouds of poison—could rend the cultural fabric of thousands of years and much of the biological fabric of billions.

The release of the nuclear genie in 1945, as Jacques-Yves Cousteau said at the United Nations in May 1976, has utterly "reshaped what we may fear, what we may dream, how we live, and how we may die." The stunning events of three decades ago still echo through the world. But we have not made over again the habits of thought and emotion that shaped our societies through hundreds of generations. The darker side of human nature has not changed. Thus Cousteau concluded:

> Despite the best efforts and intentions of the people of the United Nations, human society is too diverse, national passion too strong, human aggressiveness too deep-seated for the peaceful and the warlike atom to stay divorced for long. We cannot embrace one while abhorring the other; we must learn, if we want to live at all, to live without both.

That needed learning—its whys, hows, whos, wheres, and whens—is the burden of this final chapter. And to see how the nuclear genie came to threaten to foreclose all choices for all time, we must begin at its beginning, its first and partial emergence from the bottle that our fatal ingenuity opened. We must return—temporarily—from lofty questions of human purpose and spirit to the drier facts of science, history, and politics.

11.1 HOW THE GENIE IS GETTING OUT

In March 1946 the Acheson-Lilienthal Committee reported that if an international Atomic Development Authority did not quickly forestall national nuclear programs by gaining a monopoly over all nuclear activities with military potential, the proliferation of nuclear weapons could not be prevented. The committee concluded:

> The development of atomic energy for peaceful purposes and the development of atomic energy for bombs are in much of their course interchangeable and interdependent. From this it follows that although nations may agree not to use in bombs the atomic energy developed within their borders, the only assurance that a conversion to destructive purposes would not be made would be the pledged word and the good faith of the nation itself. This fact puts an enormous pressure upon national good faith. Indeed it creates suspicion on the part of other nations that their neighbors' pledged word will not be kept. This danger is accentuated by the unusual characteristic of atomic bombs, namely their devastating effect as a surprise weapon, that is, a weapon secretly developed and used without warning. Fear of such surprise violation of pledged word will surely break down any confidence in the pledged word of rival countries developing atomic energy if the treaty obligations and good faith of the nations are the only assurances upon which to rely.
>
> Such considerations have led to a preoccupation with systems of inspection by an international agency. . . . We have concluded unanimously that there is no prospect of security against atomic warfare in a system of international agreements to outlaw such weapons controlled *only* by a system which relies on inspection and similar police-like methods. The reasons supporting this conclusion are *not merely technical*, but primarily the inseparable political, social, and organizational problems involved in enforcing agreements between nations each free to develop atomic energy but only pledged not to use it for bombs. National rivalries in the development of atomic energy readily convertible to destructive purposes are the heart of the difficulty. So long as intrinsically dangerous activities may be carried on by nations, rivalries are inevitable and fears are engendered that place so great a pressure upon a system of international enforcement by police methods that no degree of ingenuity or technical competence could possibly hope to cope with them. . . . We are convinced that if the production of fissionable materials by national governments (or by private organizations under their control) is permitted, systems of inspection cannot by themselves be made 'effective safeguards . . . . to protect complying states

against the hazards of violations and evasions.'*** If nations may engage in this dangerous field, and only national good faith and international policing stand in the way, *the very existence of the prohibition* against the use of . . . piles to produce fissionable material suitable for bombs would tend to stimulate and encourage surreptitious evasions. ***The effort that individual states are bound to make to increase their industrial capacity and build a reserve for military potentialities will inevitably undermine any system of safeguards which permits these fundamental causes of rivalry to exist. In short, any system based on outlawing the purely military development of atomic energy and relying solely on inspection for enforcement would at the outset be surrounded by conditions which would destroy the system.[1]

Three months later, Bernard Baruch presented to the United Nations[2] a proposal to place all strategic nuclear materials and activities under effective international control with swift and certain penalties (exempt from Security Council veto) for violations—and then to destroy existing U.S. bombs. This visionary scheme became an early victim of the Cold War: the Soviet Union rejected it. And as it faded from memory, so did the reasons, so cogently argued by the Acheson-Lilienthal panel, for proposing it.

On 8 December 1953, President Eisenhower reversed a carefully constructed U.S. policy, and implicitly rejected the Acheson-Lilienthal conclusions, with his radical Atoms for Peace program.[3] "To hasten the day when fear of the atom will begin to disappear from the minds of people," he said, the nuclear weapons states, principally the United States (which had by then lost its monopoly on bombs), would contribute fissionable materials to a new International Atomic Energy Agency (IAEA), which would then redistribute them, with technical assistance, to the world, especially "to provide abundant electrical energy to the power-starved areas." In an unspecified way, this largesse would "begin to diminish the potential destructive power of the world's atomic stockpiles": "It is not enough to take

[1] C.I. Barnard et al., "A Report on the International Control of Atomic Energy," U.S. Department of State 2498 (Washington, D.C.: USGPO, 16 March 1946); reprinted in *Peaceful Nuclear Exports and Weapons Proliferation: A Compendium*, Committee on Government Operations. U.S. Senate, (Washington, D.C.: USGPO, April 1975), pp. 127–98. The detailed recommendations of the panel were unsound (and importantly so) because they relied on a tentative but incorrect assumption about "denaturing" plutonium (see Seaborg in Wohlstetter et al., note 6 *infra*, p. 70), but the broader logical framework of the Acheson-Lilienthal analysis—the part used here—remains sound.
[2] B. Baruch, statement to the United Nations Atomic Energy Commission, 14 June 1946, reprinted in *Peaceful Nuclear Exports, ibid.*, pp. 203–13, which also reprints the 1953 Eisenhower speech at pp. 214–20.
[3] *Ibid.*

this weapon out of the hands of the soldiers. It must be put into the hands of those who will know how to strip its military casing and adapt it to the arts of peace." Unfortunately, those who will know how, and will care, to do this are yet to be found, for precisely the reasons of international rivalry that the Acheson-Lilienthal panel foresaw. And though President Eisenhower assured the General Assembly that:

> The ingenuity of our scientists will provide special safe conditions under which . . . a bank of fissionable material can be made essentially immune to surprise seizure, [even without] ***a completely acceptable system of worldwide inspection and control[,]

nobody has ever been able to figure out quite what he meant.

The U.S. Atomic Energy Act of 1946 had prohibited U.S. nuclear exports, or even exchanges of information, until Congress found by joint resolution that "effective and enforceable international safeguards against the uses of atomic energy for destructive purposes" were operating. No such resolution was ever made. The Atomic Energy Act of 1954 therefore reversed this policy. It laid the basis for U.S. Agreements for Cooperation with many countries—and thus for uncorking the bottle of the nuclear genie. Dr. Victor Gilinsky of the U.S. Nuclear Regulatory Commission summarizes the consequences:

> As things turned out, weapon stockpiles were not melted down. Rather, Atoms for Peace grew into an enthusiastic worldwide distribution of the nuclear technological wealth which had been amassed in the United States. It was established policy to share our peaceful nuclear know-how as fast as it developed. . . . We worked to build up the IAEA. It did not, however, become a bank and protector of nuclear material. Instead, it provided inspection service to cover the bilateral commercial arrangements of its various members.[4]

—just the system that the Acheson-Lilienthal panel, eight years earlier, had decided would not work.

Initially the terms of the U.S. bilateral agreements, backed by monopoly power, were not grossly inadequate for the types and amounts of material transferred.

> Over the years, however, as our bilateral research and technical assistance programs fueled the mushrooming growth of commercial nuclear power

[4] V. Gilinsky, "Plutonium, Proliferation, and Policy," remarks at MIT, 1 November 1976, USNRC Release S-14-76 (in "News Releases" series 2:43, 4–7, week ending 16 November 1976). This speech is exceptionally important.

abroad, we paid more attention to the promotional function of our agreements than we did to their control provisions. This led us to see plutonium more in the light of its use in future technologies than in the present reality of its weapons potential. . . . This weighing of the relative importance of development and protection—along with the conviction that the reactor byproduct [plutonium] had an important commercial future and even that the long-term use of nuclear power depended on it—resulted in a gradual loosening of controls over dangerous material derived from U.S. exports. . . . [W]e have finally arrived at a situation in which a country can come arbitrarily close to going nuclear with our materials without violating any agreements. ***[A] desire for plutonium has been kindled by almost universally held official assumptions—our own included—that the use of plutonium is a natural, legitimate, desirable, and even indispensable result of the exploitation of nuclear power for the generation of electricity. The powerful grip of this assumption has complicated efforts to prevent this nuclear explosive material from becoming widely and freely available throughout the world.[5]

Only in late October 1976 did President Ford, echoing his election rival, call that assumption into question for the first time since 1954, at the same time criticizing national reprocessing and so bringing U.S. policy full circle back to the Acheson-Lilienthal thesis that the proliferation problem cannot be solved short of an international collaborative approach that goes far beyond international policing of national assurances. But in the intervening years the problem had been vastly complicated by the U.S.-led dissemination of nuclear materials, equipment, and knowledge to so many countries that dozens now have, or stand on the brink of, a nuclear weapons capability. The strength of U.S. leadership in this process of dissemination can hardly be overestimated: leadership as much on a political as a technical level. As recently as 1973, for example, it was at the strong personal instigation of the then U.S. Ambassador to the IAEA (currently the Director of International Government Affairs for Westinghouse) that the IAEA prepared its embarrassingly inflated[6,7] 1973-1974 Market Surveys for Nuclear Power in Developing Countries.

About 1976, largely in a slow double take after the 1974 Indian explosion, public and congressional concern in the United States, mirrored elsewhere in the Nuclear Suppliers' Group, returned belat-

---

[5] *Ibid.*
[6] R.J. Barber Associates, Inc., *LDC Nuclear Power Prospects, 1975-1990: Commercial, Economic & Security Implications*, ERDA-52 (Springfield, Virginia: NTIS, 1975).
[7] B. Johnson, "Whose Power To Choose? International Institutions and the Control of Nuclear Energy" (London: International Institute for Environment and Development [27 Mortimer St., London W.1], 1977).

edly to the level and content of thirty years earlier. Yet the policy landscape has changed profoundly since 1946, and the options still available are more subtle, difficult, and constrained. This chapter, amplifying Chapter 2.8, will argue that those options are nonetheless sufficient to reduce greatly—even to eliminate—nuclear proliferation from the world; but the medicine is strong, and the need to take it can only be understood after considering further the origin, habits, and weaknesses of the emerging nuclear genie.

Nuclear power is considered by most informed observers today to be the main driving force behind the proliferation of nuclear weapons.[8-13] A quasi-civilian industry originally intended to develop morally justifiable uses for the knowledge and equipment arising from costly military programs has expanded, with the indispensable help of military subventions, and is in its turn spawning more military programs. If, as Samuel Butler said, a chicken is an egg's idea for making eggs, then perhaps a reactor is a bomb's idea for making bombs.

The ways in which this genotype reasserts itself are many and complex. By spreading knowledge, hardware, and expectations, nuclear power greatly reduces the time and the degree of political initiative needed to convert latent to actual proliferation. Nuclear power simultaneously expands by orders of magnitude the potential scale of that proliferation, and requires the evolution of fuel cycles that increase enormously the variety of attractive proliferative paths. Nuclear power develops a self-contained institutional momentum described thus by Lowrance:

[8] A. Wohlstetter et al., "Moving Toward Life In A Nuclear Armed Crowd?" (Report to Arms Control & Disarmament Agency, U.S. State Department, ACDA/PAB-263, 22 April 1976, Pan Heuristics [1801 Avenue of the Stars, Suite 1221, Los Angeles, CA 90057]), an outstanding survey; see also the lucid and forceful summary by A. Wohlstetter, *Foreign Policy*, Winter 1976-1977, pp. 88-96, 145-79.

[9] See the report of the summer 1976 Aspen [Colorado] Institute for Humanistic Studies workshop on proliferation issues.

[10] R.N. Gardner, "Nuclear Energy and World Order: Implications for International Organizations" (Report of May 1976 Institute for Man & Science [Rensselaerville, NY] workshop on proliferation issues).

[11] Research and Policy Committee, Committee for Economic Development, "Nuclear Energy and National Security" (New York, September 1976); *cf.* L.A. Dunn & H. Kahn, "Trends in Nuclear Proliferation, 1975-1995," HI-2336-RR/3 (Croton-on-Hudson, New York: Hudson Institute, 15 May 1976); and the Flowers and Fox reports (see Chapter One, note 20) at pp. 126, 193, and 202 and at pp. 153-58 respectively.

[12] H.A. Feiveson et al., *Bull. Atom. Scient.* 32:10, 10-14 (December 1976); H.A. Feiveson & T.B. Taylor, *ibid.*, pp. 14-20; T. Greenwood et al., *Nuclear Proliferation* (New York: Council on Foreign Relations/McGraw-Hill, 1977).

[13] M. Willrich & T.B. Taylor, *Nuclear Theft: Risks and Safeguards* (Cambridge, Massachusetts: Ballinger for The Energy Policy Project, 1974).

Nuclear decisions have a certain irreversibility about them, and all have long-term consequences. Renunciation of nuclear innocence leads to a progressive hazarding of the various nuclear thresholds: securing nuclear technology, exploring the idea of weapons, making a commitment to build a first explosive device, detonating that device in a test, refining the bomb design, developing delivery capability, stockpiling more sophisitcated weapons, threatening to fire them in anger, and so, tragically, on.

The mere existence of knowledge in the laboratories tempts exploitation of it, 'just to keep the options open.' Once the options become established, pressures even from minority sources may encourage next steps. And as the nuclear weapons states have learned, once large promotionary bureaucracies are spawned they develop lives of their own, as do the specialized capital-intensive laboratories and industries that supply the technologies.[14]

Moreover, the pregnant seeds of proliferation that nuclear power sows fall not on a geopolitical desert but on ground already tilled by processes of rivalry and tension that nuclear power itself helps to activate. Because nuclear power facilities are ambiguous and have military potential, actual or planned possession of them may impel neighboring countries—the Argentinas and Pakistans of the world—to seek similar potential, if only to " 'keep *their* options open.' Around the world belligerent pairs and clusters will warily be trying to decide what moves and countermoves to make."[15]

Indeed, nuclear power is central to a whole web of hard–technology policies and perceptions that itself contributes to the global inequities, tensions, asymmetries, and frustrations that help to fuel proliferation. As Richard Falk has described a United States view of this process:

. . . [nuclear weapons are] intricately and organically connected with our prosperity, our prowess, our whole dominance of world society. To a lesser extent this same sensibility is shared. . . by other nuclear powers.

How can we keep this kind of dominance in a world of independent sovereign states? . . . [T]he only way . . . is to deny to others, or to as many others as possible, the option that we keep for ourselves.

. . . But if we're prepared to use nuclear weapons . . . , why shouldn't the same option be permitted to all those other nations that have other goals in the world? And why shouldn't those who are weaker follow the example of those who are stronger and more successful in the ways that count—according to what we've taught them? In other words, underlying the whole policy of the nuclear age is the fantastic notion that you can both promote a peaceful world and at the same time retain the domineering capacities that come from having nuclear weapons and the announced willingness to employ them. Therefore, in order to obscure the perverse logic

---

[14] W.W. Lowrance, *International Security 1*:2, 147–66 (Fall 1976).
[15] *Ibid.*

of nonproliferation, we have had to proliferate nuclear technology for so-called peaceful purposes, and that is where the Catch-22 factor comes in. In order to prevent the spread of nuclear capabilities, we've had to spread nuclear capabilities, because the problem of how to keep nuclear weapons exclusively for ourselves poses quite an impossible political puzzle. The only way we've been able to get around it is to promise a lot of technological assistance for the development of civilian nuclear programs in the non-nuclear countries. That arrangement is written right into the Nuclear Non-Proliferation Treaty.

This immense process of deception and self-deception is implicit in the effort to have the benefits of nuclear force but somehow deny them to others. . . . And then we have the audacity to lecture the Indians on their temerity in exploding a single nuclear device while in the same week that India exploded that device, the Soviet Union and the United States between them exploded seven far more destructive nuclear devices. . . .[16]

Efforts to distinguish currently "responsible" from currently "irresponsible" governments in order to proliferate nuclear power only to the former fail because they depend not only on time but on one's perspective: countries like India naturally distrust, for example, the government responsible for Hiroshima, Nagasaki, Eniwetok, Vietnam, and the continuing manufacture of roughly three new hydrogen bombs per day. Comparisons are invidious. The "perverse logic of nonproliferation" requires proliferation to all and security for none.

Some analysts, urging us to lean back and enjoy the inevitable, take the imaginative view that a world of sixty nuclear weapons states will somehow be more stable or more just than a world of approximately six. As Sir Brian Flowers replied (in his 2 December 1976 speech to the British Nuclear Energy Society), "One might as well say that because the troops have been issued with rifles, there is no harm in the rest of us carrying pistols—and there are, of course, some who sincerely believe this." Yet this view is inconsistent, since the same analysts argue that strategic deterrence has held nuclear war at bay since 1945. The theoretical basis of the deterrence doctrine crumbles with proliferation to minor nations and subnational groups (or even to an inconveniently large number of not-so-minor nations), since in a "nuclear armed crowd"[17] threats and even explosions can be anonymous, can be caused by those with little or no territory to retaliate against, and can arrive unforeseen by any of perhaps 10^{70} paths.[18] As Fred Iklé has remarked, if Americans wake up tomorrow

[16] R.A. Falk, *The Nation*, 13 March 1976, pp. 301–5. For confirmation, see K. Subrahmanyam, *Bull. Atom. Scient. 33*:2, 17–21 (February 1977).

[17] *Supra* note 8.

[18] *Hearings on the Export Reorganization Act of 1976* (19 Jan.–9 March 1976), Committee on Government Operations, U.S. Senate, USGPO 71–259, 1976, at p. 389.

and find the center of Washington gone, they might not even know who did it.

The view that nuclear power plays a major, even essential, role in the proliferation of nuclear weapons is contrary to several arguments popular among nuclear exporters (and even some IAEA public relations officials): that reactor grade plutonium, though it can be extracted clandestinely from spent fuel or stolen from the commercial fuel cycle, is unsuitable for bombs; that there are other and simpler routes to bombs that have nothing to do with nuclear power; and that the international safeguards system, notably that of the International Atomic Energy Agency, today offers and, suitably strengthened, can continue to offer ample protection. These arguments will now be considered in turn.

11.2. CAN WE HAVE PROLIFERATION WITH NUCLEAR POWER?

Whether the safeguards measures designed to prevent, deter, or detect misuses of strategic nuclear material can be defeated—a point considered in section 11.4—is irrelevant if the materials are useless once obtained. It is well known that reactor grade plutonium (exposed to high neutron fluence, hence relatively rich in isotopes of high spontaneous fission rate such as ^{240}Pu) is not ideally suited for making explosives. This is because the flux of spontaneous fission and (α, n) neutrons with Poisson statistics makes it likely that a chain reaction will be initiated early enough to cause disassembly before maximum supercriticality has been achieved. The result is a highly uncertain yield and, ordinarily, rather low average efficiency and yield. This result has been widely misconstrued as implying that such a bomb is innocuous, or particularly hazardous to its maker, or militarily insignificant. None of these interpretations is correct. In amplifying this point the intention is not to help the malicious (who will learn nothing new here) but to contribute to sound policy, on the principle that the only thing more dangerous than talking about these repellent matters is not talking about them—that is, letting policy rest on wishful thinking that is falsified empirically.

It is now conceded by all but the uninformed that reactor grade plutonium can credibly be used by terrorists, on the basis only of the extensive information in the open literature, to make crude but convincing bombs—of unpredictable yield (typically of order 0.1–1 kiloton), but still sufficient to wipe out a government, breach the

---

[19] *Supra* notes 8 and 13.
[20] D.B. Hall, "The Adaptability of Fissile Materials to Nuclear Explosives," October 1971 manuscript (out of print); a slightly revised version is reprinted

containment of a nuclear facility, etc.[19-22] Some official denials persist[23] but are so riddled with simple technical errors as to enjoy little credence. (For example, reduction of oxide to metal, foundry and machining operations, and use of a barely subcritical mass are not necessary as claimed.[24]) A forthcoming study by the U.S. Office of Technology Assessment should help to consign such misleading statements to deserved obscurity.[25]

The dangers incurred by amateur bombmakers, though significant, have been greatly exaggerated in some quarters. Predetonation, the process described above for crude designs using reactor grade plutonium, means "premature" on a time scale of the order of a millionth of a second; it does not mean that the bomb is more likely, because of the high neutron background in reactor grade plutonium, to go off while being assembled or transported. (Reactor grade plutonium does give off more heat, which must be allowed for.) A generous safety margin of subcriticality can be used, particularly; with nonmetallic cores. The procedures for handling strategic materials and high explosives safely are widely known and are described in many textbooks. The optional processing operations are comparable in risk to those of other operations, such as heroin synthesis, that have long been carried on successfully in secret criminal laboratories. Indeed, as Walter Patterson notes, some nuclear advocates inconsistently claim that the same toxic materials that would, they say, be almost surely fatal to malicious people trying to work with them would nonetheless be all but harmless to the public if deliberately dispersed! And of course it is not necessary to go to the trouble and risk of actually making a bomb if instead one can make a credible threat of having done so.

Unfortunately, in the fuss over what terrorists can or cannot do, many people have forgotten that by using sophisticated design techniques, unlikely to be available to amateurs, reactor grade plutonium can be made into a rather high yield and predictable bomb. Such

---

in R.B. Leachman & P. Althoff, eds., *Preventing Nuclear Theft: Guidelines for Industry and Government* (New York: Praeger, 1972), pp. 275-83.

[21] J.C. Mark, "Nuclear Weapons Technology," and ensuing discussion, B.T. Feld et al., eds., *Impact of New Technologies on the Arms Race* (Cambridge, Massachusetts: MIT Press, 1971), pp. 133-38.

[22] P. Jauho & J. Virtamo, "The Effect of Peaceful Use of Atomic Energy upon Nuclear Proliferation" (Helsinki Arms Control Seminar, Finnish Academy of Arts and Letters, June 1973).

[23] A.D. Starbird, in *Hearings, supra* note 18, at 449-51. Such apologists sometimes cite the effective "denaturing" of plutonium as a further option, but effective denaturing is categorically impossible, since sources such as ^{252}Cf can be readily stripped out by ion exchange.

[24] *Ibid.*

[25] *Supra* note 8.

people have therefore assumed that governments wishing to make bombs will juggle civilian fuel cycles (especially those with on-load refueling and good neutron economy, such as CANDU) for low burn-up, either evading international safeguards or discharging fuel prematurely because of real or claimed fuel damage—or, more likely, will unambiguously announce their intentions to reconnaissance satellites and other intelligence media by making clandestine "dedicated facilities" specifically to produce weapons grade plutonium.

Careful consideration of this problem leads to a different conclusion: that the design techniques likely to be available to small, modestly sophisticated governmental weapons programs—essentially the same techniques that such programs would be likely to use anyhow—would permit reactor grade plutonium to be made directly into bombs of substantial military utility. Yields could be made, if not wholly predictable, at least in the kiloton range, with a higher expected value. Even higher predictability and yield could be obtained with nuclear testing. This conclusion, shared by many professionals, has been confirmed by Commissioner Gilinsky:

> . . .so far as reactor-grade plutonium is concerned, the fact is that it is possible to use this material for nuclear warheads at all levels of technical sophistication. In other words, countries less advanced than the major industrial powers but nevertheless possessing nuclear power programs can make very respectable weapons. And, I might add, these are the very countries whose names turn up in every discussion of proliferation. Of course, when reactor grade plutonium is used there may be a penalty in performance that is considerable or insignificant, depending on the weapon design. But whatever we might once have thought, we now know that even simple designs, albeit with some uncertainties in yield, can serve as effective, highly powerful weapons—reliably in the kiloton range.
>
> Unfortunately, the IAEA itself does not seem to have been entirely clear about this, and a number of non-nuclear states have raised questions about it. It is vitally important to serious attempts to stop further proliferation that any genuine confusion or misapprehension abroad about whether effective nuclear weapons can be manufactured with plutonium from power reactors be cleared up promptly. Such misapprehensions do exist: I encountered them myself a few weeks ago in meetings with high officials in Europe. This is bad enough, but it is deeply disturbing to encounter irresponsible encouragement of such notions at home. For it will never be possible to discourage national reprocessing if it is believed the United States is faking the dangers.[26]

Thus, if we assume that reactor grade plutonium is not militarily

[26] *Supra* note 4.

significant, a country that knows better can divert this material from its own or another country's plutonium fuel cycle—or, at the risk of signaling its intentions, build a clandestine reprocessing plant to take diverted spent fuel—without in either case building a production reactor or altering the "normal" burnup[27] of a power or research reactor. For a bomb performance penalty that can, with suitable and attainable design, be made "insignificant," a country can therefore make military plutonium without unambiguously appearing to be doing so. Retaining the ambiguity, large scale, technical assistance, and generous credit terms of the civilian program may be politically and economically attractive enough to make it well worth the extra few years' construction time.

The military potential of reactor grade plutonium is not an academic nicety. It places a very different complexion on civilian nuclear power programs, for these are expected to accumulate extremely large amounts of reactor grade plutonium very rapidly in many countries.[27] These projected civilian stocks now appear in a different light—a problem much more imminent, and orders of magnitude larger, than the stocks that might otherwise be made clandestinely in dedicated production reactors. A 1000 MW(e) power reactor, coupled either to a cheap, dirty, secret reprocessing plant or to an overt commercial reprocessing plant in which losses below about 1 percent of throughput are lost in statistical noise,[28] is merely a large-scale, inherently ambiguous military-civilian plutonium production reactor making co-product electricity: no more, no less.

Moreover, such a reactor brings with it skills, knowledge, and equipment that can be readily turned to military programs. It entails research and test reactors that are often well suited to military production or research. If intended as more than a flash in the pan, a power reactor also brings a thrust toward reprocessing, plutonium recycle, and plutonium breeding, with strategic material becoming an indispensible item of large-scale international commerce. Breeder fuel cycles entail nearly an order of magnitude larger plutonium flows than equivalent thermal reactor fuel cycles, and for economic reasons are likely to keep blanket plutonium (of low ^{240}Pu content and thus more convenient for high grade explosives) separate from high burnup core plutonium throughout reprocessing. Thus an interim uranium economy cannot be considered in isolation from successive and increasingly proliferative forms of plutonium economy.[29] The vul-

[27] *Supra* note 8, especially the *Foreign Policy* paper.
[28] *Supra* note 18 at 1016.
[29] *Supra* notes 8, 12, and 13.

nerability of any of these fuel cycles to governmental or subgovern-
mental diversion will be considered in section 11.4.

11.3. CAN WE HAVE PROLIFERATION WITHOUT NUCLEAR POWER?

As advocates of nuclear exports are fond of pointing out, countries
desiring bombs can obtain them without having civilian nuclear
power programs, though in such a case the necessary actions, if
detected, are generally unambiguous and therefore carry a higher
political cost. These actions must also, unlike civilian programs, use
the bombmakers' own money and expertise.

One possible route to bombs is buying or stealing military bombs
or components. At least one country (Libya) is widely thought to
be in the market. From limited discussion in the open literature, it
appears that sufficiently determined thieves could obtain military
bombs.[30] U.S. and, presumably, other Western military bombs
require elaborate arming devices and procedures before they can
explode. How far these devices and procedures are available to un-
authorized past or present military personnel, large numbers of whom
must have been appropriately trained, is not clear. How far the
bombs are designed to be tamperproof is also not clear. In principle
they should be designed to disperse their cores (by high explosive but
not nuclear detonation) if tampered with, but such a precaution may
be considered an operational hazard and is probably not taken. There
is a widespread impression in the open literature that safety and arming devices are probably less elaborate in non-Western than in U.S.
bombs: the research and development that went into U.S. safety
devices was difficult and costly. In summary, it may be possible for
stolen military bombs to be exploded or for their cores to be recovered and reused. This is a significant hazard to which there seems no
really effective response other than universal strategic and tactical
nuclear arms reduction and, ultimately, elimination. As will be discussed in section 11.5, this laudable long-term aim is intimately
related to civilian denuclearization.

Another route would be for a country, either at a moment of crisis
or more methodically and clandestinely, to make into bombs research
material imported from the U.S. or other countries. Table 11-1
shows a published summary[31] —said by a knowledgeable IAEA offi-

---

[30] *Supra* note 1 at 499–503 (statements of Senators Pastore and Baker); F.
Barnaby, *New Scientist 66*:494 (29 May 1975).
[31] *Supra* note 18 at 18–19, 813.

Table 11-1.   Gross U.S. Exports of Strategic Nuclear
Materials Up To 31 March 1976[a]

Country	kg ^{235}U in U enriched > 20 percent	kg $(Pu + {}^{233}U)$[b]
Argentina	71.8	0.2
Australia	30.2	7.3
Austria	31.1	0.3
Belgium[c]	301.3	18.2
Brazil	26.9	0.1
Canada	1125.9	4.7
Colombia	2.8	—
Denmark	34.8	0.1
Finland	2.9	0.1
France	4465.2	14.4
Germany (Federal Republic)	4061.1	795.0
Greece	12.1	0.1
Indonesia	2.3	—
Iran	5.2	0.1
Israel	17.0	0.6
Italy	479.6	129.1
Japan	960.9	116.1
Korea (South)	2.7	—
Mexico	—	0.2
Netherlands	225.8	0.5
Norway	2.2	1.1
Pakistan	5.2	0.1
Phillipines	8.8	—
Portugal	13.1	—
South Africa	54.4	0.2
Spain	74.3	—
Sweden	219.9	1.1
Switzerland	20.5	1.5
Taiwan	18.0	0.7
Thailand	4.8	0.1
Turkey	4.8	0.3
United Kingdom	2367.0	22.4
Venezuela	4.9	—
Vietnam (South)	2.4	0.1
Uraguay	—	0.1
Yugoslavia	3.4	—
Zaire	2.0	—

[a]Not net of returns, and therefore not all necessarily in the stated country at any one time.

[b]Fissile content of Pu not available, nor any other strategic materials exported, such as ^{237}Np.

[c]Including Eurochemic.

cial to be incomplete[32] —of the large amounts of strategic material already exported from the U.S. alone. It appears that, among non-nuclear weapons countries, Australia, Belgium, Canada, West Germany, Italy, Japan, and perhaps Norway, Sweden, and Switzerland have been sent by the U.S. (gross, and not necessarily all at the same time) enough cumulative extracted Pu and ^{233}U to make one or more bombs. Moreover, though published information does not say how far above 20 percent the uranium shipments have been enriched in ^{235}U, there is a strong presumption that many of those shipments—reportedly some five tons[33] —would have been highly enriched and therefore a straightforward weapons material giving, perhaps, a weapons capability to at least half the countries listed. (The critical mass of ^{235}U in a sphere of 93.5 percent-enriched metal in a thick beryllium reflector is less than 10 kg at unit density, and can be very substantially reduced by implosion.) Inventories of research material instantaneously present in recipient countres are generally not published, but it is known[34] that in March 1975 South Africa, which is said to have an indigenous enriched uranium weapons program, held 7 kg and had on order 39 kg more of unirradiated U.S.-supplied uranium, enriched to 90 percent, as fuel for the Safari reactor, in a physical form readily convertible to military use. Countries not receiving, or not having on hand at a given time, enough strategic material to make into a bomb would still have enough to do useful design studies and materials experiments with, and perhaps enough to use or sell as part of the basis of a threat.

A particularly common source of fully enriched uranium in easily extracted form is fresh cores for certain types of small research reactors. It is possible without impropriety for a country having such a reactor to shop for fresh cores from the four vendor countries, rather like a registered drug addict collecting a sheaf of prescriptions from different doctors. A sensitive country was recently discovered (by accident, and just in time) to be doing exactly that, and several other broadly similar incidents have been discussed in general terms within the safeguards community.

If such convenient material is not available by gift, loan, or pur-

---

[32] The totals are in some cases smaller than those shown as U.S. exports under IAEA safeguards in IAEA annual reports. At 31 December 1975 a total of 634 kg of Pu had been exported from the U.S. to non-NPT countries under IAEA safeguards but only 18 to NPT countries (plus 1060 to Euratom countries, 602 to the U.K., and less than 1 kg each to Uraguay and Iraq).

[33] *Supra* note 8.

[34] *Hearings on the Export Reorganization Act—1975* (24 April–1 May 1975), Committee on Government Operations, U.S. Senate (Washington, D.C.: USGPO 55-430, 1975), pp. 220–22.

chase—or by theft, as physical security is generally even more lax outside the U.S. than inside—the patiently diverted output of a research reactor can be reprocessed to extract plutonium. A suitable small-scale reprocessing plant, not intended to be commercially viable nor particularly clean or safe, might be built[35] in a year or two for $1-3 million at a scale sufficient to yield several bombs per year. Laboratory scale operations at a smaller scale are much simpler and cheaper.

As will be argued in section 11.4 below, any safeguards on reactor output, or on fresh cores or research material, can be renounced or evaded. On the other hand, a country that was not adopting nuclear power, in a world that was phasing it out, could hardly justify operating (much less building anew) research reactors and similar facilities. Thus a policy of general denuclearization would isolate unambiguously any countries intending to continue with nuclear research for presumably military purposes. To the extent that the phasing out of research reactors made it impracticable to use for medical and industrial purposes a very limited number of short-half-life isotopes that could not be made in one or two multinational research isotope centers, that small price would have to be paid to isolate as presumptively military any residual research reactors.

A further alternative route to bombs is to build a natural uranium, graphite-moderated (or, conceivably, heavy-water-moderated) production reactor using designs fully described in the open literature. This would take a few years and cost perhaps $13-26 million[36] (less by some estimates). It could be made much more difficult and noticeable by better controls on such strategic materials as nuclear-grade graphite, zirconium, and beryllium, as well as on uranium, deuterium, and nuclear instrumentation. Intelligence machinery could be concentrated on such materials in a world in which such items were no longer being supplied by many vendors in many countries as a major item of world commerce. Again, denuclearization would make those activities unambiguously military in presumed purpose, and would offer psychological deterrents and disincentives while reducing nuclear threats by all countries to all others—threats that now help to motivate military nuclear programs.

---

[35] W. Epstein, *The Last Chance, Nuclear Proliferation and Arms Control* (Riverside, New Jersey: Free Press, 1976). T.B. Taylor thinks the cost might be lower. See also J.R. Lamarsh, "On the Extraction of Plutonium from Reactor Fuel by Small and/or Developing Countries," 19 July 1976, report for U.S. Congressional Research Service.

[36] J.R. Lamarsh, "On the Construction of Plutonium-Producing Reactors by Small and/or Developing Nations," QC 170 Gen., 30 April 1976, report for U.S. Congressional Research Service, reprinted in *Hearings, supra* note 18 at 1326-55.

A further possible route to self-sufficiency in bombs is uranium enrichment. Gaseous diffusion and nozzle enrichment have been and may be used, respectively, for this purpose, but both require of order 10^3 stages and are thus conspicuous as well as very costly in money and energy. It does not appear to be physically possible to scale down these processes (or ion exchange methods, if these work at all) to cheap, simple, modular, inconspicuous devices.

The same does not appear to be true of two other enrichment methods. Ultracentrifuge technology in its prototype commercial form is a very high technology that, unless deliberately exported, is unlikely to become available to countries less technically sophisticated than, say, the Netherlands. But according to one expert, "the construction of a set of crude centrifuges does not require unusual skills as long as efficiency and commercial competition are not the point."[37] Apparently this contention was empirically verified in a test in a period of about three weeks.[38] The technology is modular and can slowly produce bomb material from tens of units or perhaps fewer; thus a clever technician should be able, in due course, to make a bomb from several tons of natural uranium, which is widely distributed in the earth's crust and in seawater (and hundreds of tons of which are used each year for nonnuclear purposes such as ceramics, though that would not necessarily continue to be the case in a denuclearized world). Uranium ore would thus become a strategic commodity and we would be back to the Baruch plan.

The same problem arises, albeit in more speculative form, with laser enrichment.[39] If this experimental process turns out to have tractable engineering and chemistry on a crude pilot scale, highly enriched uranium could be made in a single pass through a device that might be as cheap and simple as an array of ultracentrifuges. A recent review of the physical and engineering basis of laser enrichment[40] suggests that the technology is unlikely to be easy, but nobody can rule out the possibility that it might become accessible to a wide range of proliferators.[41] If there is any chance—as there appears to be—that some clever form of laser enrichment will be relatively easy, then our best protection would be to have a fission-free economy in which uranium is neither ordinarily mined, nor extracted

[37] W. Häfele, "Non-Proliferation of Nuclear Weapons," typescript (Laxenburg, Austria: IIASA, December 1976), p. 10.

[38] W. Häfele, personal communication, 16 February 1977.

[39] B.M. Casper, *Bull. Atom. Scient. 33*:1, 28–41 (January 1977).

[40] A.S. Krass, "Laser Enrichment of Uranium: The Proliferation Connection," PU/CES 42, Center for Environmental Studies, Princeton University, 15 December 1976; submitted to *Science*.

[41] G.W. Rathjens takes a particularly gloomy view in *Hearings, supra* note 18 at 106.

as a by-product (e.g., in phosphate mining), nor traded in international commerce. The quantities needed to make a bomb would then no longer be an undetectable fraction detachable from commercial flows, and any interest in uranium would be presumptively military. If there turned out to be no worryingly simple form of laser enrichment, we would still have played safe and at the same time inhibited centrifuge builders, while the world order implications of denuclearization would have reduced the temptation to pursue either the laser or the centrifuge route to bombmaking. As an interim precaution, research in laser enrichment (and, for analogous reasons, in laser fusion) could and should be abandoned in order not to increase the stock of knowledge whose dangers far outweigh its potential benefits.

Several other routes to bombs can be imagined. Large charged-particle accelerators, not now safeguarded, can be used to make substantial amounts of strategic material,[42] though such machines are scarce and expensive enough not to constitute a serious global safeguards problem. Electromagnetic isotope enrichment probably falls in the same category. The copious fast neutron flux from thermonuclear or near thermonuclear controlled fusion devices would offer a great temptation for fission breeding, and, as mentioned in Chapter Two, note 20, that is one of several excellent reasons not to build such devices. And in the long run, knowledge of how to recover and use plutonium might have a longer half-life than the principal fission products with which plutonium is mixed in irradiated nuclear fuel and reprocessing wastes, so such materials must be disposed of in inaccessible places, and we ought not to make more of them than we already have. Likewise, the transuranic isotopes which, while not fissionable by thermal neutrons, can nonetheless be used to make fast critical assemblies—neptunium-237 is the most important—should immediately be placed under the same restrictions of production and use that apply to other bomb materials today, and again we should not further increase our stocks of them.

In summary, then, the answer to the question at the head of this section is "yes, but." Yes, we can have proliferation without nuclear power; but it can be made much more difficult and easily detectable by removing its civilian "cover," it can be made unambiguously military if detected, and the motivation for it can be largely removed under the type of program of world denuclearization discussed in sections 2.8 and 11.5.

[42] T.B. Taylor, in "Oversight Hearings on Nuclear Energy—International Proliferation of Nuclear Technology," Subcommittee on Energy and the Environment, Committee on Interior & Insular Affairs, USHR, serial 94–16, Part III (Washington, D.C.: USGPO, 1975), p. 74.

11.4. CAN WE HAVE NUCLEAR POWER
WITHOUT PROLIFERATION?

Since, as noted earlier, strategic materials usable in both crude and sophisticated bombs are produced by all power reactors (and are required as fuel by plutonium breeders[43] and recyclers[44] and by current designs of the now defunct high temperature gas-cooled reactors), nuclear power without proliferation entails technical and political arrangements to prevent, deter, and detect in good time the covert or overt theft or diversion of strategic materials—and then to recover the materials, the most difficult step. For the reasons set out below, even the prerecovery parts of this process seem impossible in principle.

Strategic materials may be removed from civilian uses by overt or covert theft or by governmental (or factional or industrial[45]) diversion[46,47]—the first breaching physical security measures, the second breaching any applicable international safeguards through actions within the purview of national sovereignty or equivalent de facto control. Governmental diversion may, of course, be advantageously disguised as private theft.

Resisting overt theft entails armed force and added police powers. The second may prove intolerable and the first inadequate, since the level of force and sophistication used by the thieves can be increased within wide limits. Colocation of fuel cycle facilities or reactors, the use of heavy casks and secure vehicles, alarms, guard forces, and (at costs and occupational risks judged by USNRC contractors in the current Special Safeguards Study to be excessive) "spiking" of strategic materials with hard gamma emitters can all make theft more difficult—but not impossible. The permanent weaknesses of any pro-

---

[43] Feiveson and Taylor, *supra* note 12, suggest a thorium cycle in which plutonium is not an item of international commerce but is instead burned in special reactors sited at multinational reprocessing centers. As a partial, theoretical technical fix it is not bad—though many details are not yet clear—and it is likely to fall instead on the political problems of multinational reprocessing (*q.v. infra*) and on the generic objections to any type of nuclear economy.

[44] It is not particularly difficult to separate PuO_2 from fresh mixed oxide fuel, though a larger quantity of such fuel must of course be stolen (see note 13, *supra*). The most vulnerable point in the cycle would be that at which Pu has not yet been recombined into mixed oxide and can be embezzled or, perhaps, seized. Coprecipitation of PuO_2 and UO_2 is possible in principle and would require thieves to resort to reseparation of PuO_2 from mixed oxides.

[45] Two alleged incidents come to mind: the fairly reliable reports cited in note 87, *infra*, and the more controversial ones by H. Kohn in *Rolling Stone*, 13 January 1977, pp. 30–39.

[46] *Supra* note 13.

[47] D.M. Rosenbaum et al., "A Special Safeguards Study," internal task force report to Director of Reactor Licensing, USAEC, 1974; reprinted 120 *Congr. Rec.* S6621-30 (30 April 1974).

tective schemes will be human corruptibility and carelessness,[48-50] laxity of precautions in some countries (physical security is now inadequate[51] in the U.S. and in general is grossly inadequate elsewhere despite the known risks), and the proven ability of criminals and terrorists to mount meticulously planned paramilitary operations to obtain valuable materials.[52] Most security specialists concede that it is impossible to protect anything from people who wish intensely enough to obtain it. The failure of the international community to halt aircraft hijackings, fiscal embezzlement, bank robberies, and the black market in heroin stand as disquieting reminders: the more so because the open market price of ^{239}Pu is comparable to the black market price of heroin, so the black market price of ^{239}Pu might be very much higher. Some observers believe that such a black market in strategic materials may already exist.[53]

Resisting covert theft is even more difficult. It entails extensive secret policing with serious civil liberties implications.[54] Covert theft can be made undetectable in principle, since the precision of inventory assay (now nearly 1 percent) will never be good enough to reveal gradual but cumulatively significant thefts, and since complex materials accountancy methods can be readily misled by dishonest insiders[55]—especially the very complex computerized methods proposed for dynamic materials accountancy,[56] which add computer

[48] *Ibid.*

[49] See proceedings of the annual meetings of the Institute of Nuclear Materials Management, 1969- .

[50] Comptroller General of the U.S., B-164105, U.S. General Accounting Office, 7 November 1973 and 12 April 1974, and EMD-76-3a, 1976; R. Gillette, *Science 182*: 1112 (1973); *Nature 246*:241 (1973); C.H. Builder, memorandum reprinted in "Oversight Hearings," *supra* note 42, Part VI, pp. 214-16 (*cf.* discussion at pp. 253-58); "Joint ERDA-NRC Task Force on Safeguards (U) Final Report," NUREG-0095/ERDA 77-34, July 1976, cogently discussed by T.B. Cochran and A.R. Tamplin, "Nuclear Weapons Proliferation—The State Threat and the Non-State Adversary" (Washington, D.C.: Natural Resources Defense Council, 2 March 1977).

[51] *Supra* notes 47 and 50.

[52] S. Burnham, ed., "The Threat to Licensed Nuclear Facilities," MITRE MTR-7022, September 1975, pp. 56-59; T.B. Taylor, in M. Willrich, ed., *International Safeguards and Nuclear Industry* (Baltimore and London: Johns Hopkins University Press for American Society of International Law, 1973), pp. 188-91.

[53] *Supra* note 35.

[54] See Chapter Two, note 34 (the Ayres paper is reprinted in "Oversight Hearings," *supra* note 42, Part VI, pp. 259-333); *ibid.*, p. 249; Chapter Nine, note 21; K.P. O'Connor, "Impact of the Proposal to Mine Uranium in Australia on Civil Liberties," submission to Ranger Uranium Environmental Inquiry (Melbourne hearings), March 1976; J.G. Speth et al., *Bull. Atom. Scient. 30*: 9, 15 (November 1974); *Der Spiegel*, 28 February 1977 and following issues, for accounts of the Traube affair.

[55] *Supra* note 47.

[56] J. Googin, quoted in *121 Congr. Rec.* (11 March 1975), pp. S3619-20.

fraud to the scope for deceiving the instruments. Covert thefts large enough to be detected at all can be made undetectable within a time longer than that required to remove the material to unknown places from which recovery might be difficult or impossible and attempts at recovery would certainly be repressive[57] : in short, the certain and timely detection of covert theft can be made impossible in principle.

The fundamental reason that nuclear theft cannot be prevented is that people and human institutions are imperfect. No arrangement, however good it looks on paper, and however competently and devotedly it is established, is proof against boredom, laxity, or corruption. Perhaps a few historical examples relevant to nuclear safeguards might illustrate this conclusion common to everday experience:

- In 1973 the former security director of the U.S. Atomic Energy Commission was sentenced to three years' probation; he had borrowed $239,000 from fellow AEC employees, spent much of it on racetrack gambling, and failed to repay over $170,000.[58]
- A security guard at the former Kerr McGee plutonium fuel fabrication plant in Oklahoma was arrested in connection with an armed robbery of a loan company in which a woman was shot; he was found to be a convicted and paroled armed robber hired under a false name.[59]
- Each year about 3 percent of the approximately 120,000 people working with U.S. nuclear weapons are relieved of duty owing to drug use, mental instability, or other security risks.[60]
- In 1974 several tons of unclassified metal were stolen from the nuclear submarine refitting docks at Rosyth, Scotland, apparently through a conspiracy of dockyard employees.[61] Nuclear submarine fuel, present at the same docks, is fully enriched uranium and thus bomb material.
- In 1976 more than a ton of lead shielding was reported missing from Lawrence Livermore Laboratory, a U.S. nuclear weapons design facility.[62]

[57] *Supra* note 54.

[58] M. Satchell, "Ex-AEC Aide Put on Probation," *Washington Star-News*, 21 February 1973.

[59] "Silkwood, Initiative Campaigns Heating Up," *Nuclear News*, November 1975.

[60] AP, *Observer* (London), 27 January 1974, p. 1; UPI, *Intl. Her. Trib.*, 14–15 December 1974, p. 3; *LA Times*, 27 January 1974. All the stories were based on congressional testimony by a senior federal official.

[61] *Times* (London), 16 July 1974, p. 1; *cf. Hansard* 276-77, 19 July 1974, House of Commons written answer.

[62] KFWB Radio (Los Angeles), quoted by L.D. DeNike, "Oversight Hearings," *supra* note 42, part VI, p. 22.

- Reports persist that materials accountancy records for highly enriched uranium may have been fraudulently manipulated at the Erwin, Tennessee, plant in order to accumulate a surplus with which to cover possible future losses.[63]
- An analytic laboratory used by the Japanese nuclear industry to monitor effluents was shut down by the government for falsifying and fabricating test results.[64]
- During the 1967 Cultural Revolution the military commander of Sinkiang Province reportedly threatened to take over the nuclear base there.[65]
- French scientists testing a bomb in the Algerian Sahara apparently had to destroy it hurriedly in 1961 lest it fall into the hands of rebellious French generals led by Maurice Challe.[66]
- A smuggling ring was found in 1974 to have been selling uranium stolen from an Indian plant to Chinese or Pakistani agents in Katmandu.[67]
- Senior Italian military and intelligence officers were arrested in 1974 for allegedly plotting a right-wing coup in which public panic was to be generated by adding to water supplies some radioactive materials to be stolen from a research center.[68]

Nor is the fatal combination of personal fallibility and institutional rigidity peculiar to China, Algeria, or Italy. A Canadian scientist, for example, has described an official program he operated to teach plutonium metallurgy to visiting Indian scientists more than a decade before India exploded a bomb made of plutonium from a Canadian reactor (plus U.S. heavy water and, reportedly, other contributions from France and West Germany). The Indian visitors said they planned to go home and set up their own metallurgical laboratory— entirely for peaceful purposes, of course—but they also displayed an intense interest in some subjects with no known civilian applications. The Canadians running the program knew just what was happening, but it appears that the Canadian export authorities did not care to act on that knowledge.

Indeed, according to an authoritative study by Gowing[69] and a

[63] *Ibid.*, p. 193; and B. Newman, *The Nation*, 23 October 1976, p. 396. *Cf. Hearings, supra* note 18 at 695–96.

[64] *Nucl. Eng. Intl.* 19:214, 123 (March 1974).

[65] D.G. Brennan, *Arms Control & Disarmament* 1:59–60 (Pergamon, UK) (1968).

[66] *Ibid.*; senior French officials have privately confirmed the substance.

[67] *Times* (London), 2 May 1974, p. 8; *ibid.*, 8 October 1974.

[68] Group W radio news (and other wire service reports), 25 October 1974, reporting a statement to Parliament by Defense Minister Andreotti.

[69] M. Gowing, *Independence and Deterrence*, 2 vols. (London: Macmillan, 1974).

somewhat more oblique remark by Scheinman,[70] both Britain and France, respectively, were unable to subject their own original nuclear weapons programs to political control or formal decision before those programs were virtually completed. That this should happen in the countries with perhaps the greatest experience of exercising political control suggests that it could happen anywhere, even with a "normal" government—let alone on the occasions Hans Bethe referred to[71] when he approvingly quoted Enrico Fermi's remark: "Every country has a mad dictator every few centuries." This imperfection of governments places fundamental limits on the efficacy of international safeguards and domestic physical security, both of which depend on the effective exercise of national sovereignty and good faith. Those limits are especially important because the materials relevant to safeguards are the same ones that make nuclear waste management so intractable: materials that last many times as long as the interval from the Neolithic to the present. Thus the technical and political measures intended to prevent endemic nuclear violence and coercion must last not merely for the usual lifetime of treaties or nations but for the lifetime of the materials themselves. Even the most generous enthusiasts of effective international safeguards would hardly claim that this is plausible. As Falk remarks:

> Considering the terrorism and the upheaval that one sees everywhere in the world today, not to mention the relatively short duration of powerful empires throughout human history, the notion of creating eternal vigilance and permanent social institutions strikes me as the most utopian suggestion I've ever heard. Think how proud we [in the U.S.] are to be 200 years old, and already we've had one Civil War and participated in two World Wars. Even to be in favor of something one calls a Faustian bargain suggests either a rather weak capacity for literary extrapolation, or else a pathological compulsion.[72]

Even within our few decades of experience with the international arrangements intended to facilitate Atoms for Peace without reawakening Atoms for War, the rivalry of nations and the fallibility of people have proved the Acheson-Lilienthal panel right. The whole network of bilateral and multilateral safeguards, including those of the IAEA and the Non-Proliferation Treaty, is deeply flawed, from

---

[70] L. Scheinman, *Atomic Policy in France Under the Fourth Republic*, (Prineton, New Jersey: Princeton University Press, 1965), as quoted in Wohlstetter et al., *supra* note 8 at 44–45.

[71] *Hearings, supra* note 18 at 23. Likewise, the Flowers report (see Chapter One, note 20) at para. 167 saw "no reason to trust in the stability of any nation of any political persuasion for centuries ahead."

[72] *Supra* note 16.

the start "surrounded by conditions which would destroy the system"[73]: so much so that as Wohlstetter points out,[74] many countries today, without breaking any agreement, can come closer to having a bomb ready for use than the U.S. was in 1947. There is a vast literature on these inherent flaws.[75-78] The problems of nonadherence, freedom to withdraw, lack of sanctions, inadequacy of inspectors for "their immense and dreary task,"[79] inadequacy of inspection safeguards in principle, safeguards confidentiality, ability of the inspected country to deceive the inspector in a dozen ways or to make his job impossible in a dozen more, freedom to do all sorts of nuclear weapons R&D short of actually assembling strategic material into a bomb, loopholes in existing agreements, and so on are all well known.

Lately some analysts have started to say so. Lowrance, for example, stresses, as the Acheson-Lilienthal panel did, the need to rely on the good faith of precisely the countries that have the strongest incentives not to exhibit it:

> Reassuring repetition of the word "safeguards" notwithstanding, technical safeguards are no more than procedures for conducting moderately close inventories and inspections—their only power being deterrence by threat of detection. Safeguards are indeed essential, but they are by no means such a deterrent to rash action that all the other forms of international politics can be ignored. Far from it. In fact, in the extreme the glib assurances of safeguards may simply serve as a Trojan Horse allowing dangerous technologies to be conveyed under a guise of benignity. Safeguards agreements can be abrogated on short notice. Extremely serious acts can be accomplished before the United Nations Security Council can take action, and there is not much in the Council's history to inspire confidence in its powers of enforcement. Sanctions, political and economic punishments, are often referred to, but, again regrettably, history shows that these have always proven easier to discuss than to enforce.[80]

[73] *Supra* note 1.

[74] *Supra* note 8.

[75] *Supra* notes 4 and 6-8.

[76] A.W. Marks, ed., *NPT: Paradoxes and Problems* (Washington, D.C.: Arms Control Association/Carnegie Endowment for International Peace, 1975).

[77] See the Fox report (Chapter One, note 20) and the April 1976 submission thereto by Dr. J.A. Camilleri (Dept. of Politics, Latrobe University, Melbourne), "International Implications of Uranium Exports."

[78] Comptroller General of the U.S., "The Role of the International Atomic Energy Agency in Safeguarding Nuclear Material," report to the Committee on International Relations, U.S. House of Representatives, ID-75-65, 3 July 1975, excerpted in *Hearings, supra* note 18 at 922-42; B. O'Leary, "Critique of IAEA Safeguards," *ibid.*, pp. 1452-63.

[79] *Supra* note 1.

[80] *Supra* note 14. The Fox report (*supra* note 77) is at least as outspoken.

Commissioner Gilinsky, too, emphasizes the limits of policing independent national programs by international inspection:

> While it is a cliche of the inspection trade that diversion cannot be *prevented* by inspection safeguards, there is nevertheless a general human tendency to relax and assume protection once they are in place. Calling the inspections "safeguards" contributes to the illusion. . . . Of course, since separated plutonium can be turned to weapons use in a matter of days, it is difficult to see how inspection safeguards over reprocessing and separated plutonium can give any protection. This is in fact why we don't want to see these facilities spread around, although we have been reluctant to say so squarely because this might "undermine" the IAEA.[81]

A fair summary of existing international safeguards, then, would seem to be that they are probably better than nothing—by an unknown amount—provided that their inherent limitations are clearly understood. This is clearly not the case in many influential quarters.

For example, many people still do not appreciate that the IAEA not only does not undertake to prevent nuclear theft or diversion (physical security being the responsibility of sovereign governments), but does not even undertake to detect thefts in good time—that is, in time to try to recover the material or at least to take precautions against its imminent use. The period needed between stealing, say, reactor grade plutonium dioxide and making it into a convincing bomb is certainly as short as days,[82,83] and a proper basis for planning would be to assume the interval is only twelve hours[84]—itself not necessarily a conservatively low figure. Even an inspector who happened to be on the spot and to detect the theft instantly could not be expected to confirm that he really saw it, notify the Inspector-General, and through him activate the ponderous machinery of the IAEA's Board of Governors and the U.N. Security Council before a bomb had been made and exploded. Moreover, the thefts or losses of IAEA-safeguarded material that have been detected[85] —so far apparently not in strategic quantities—do not inspire confidence in the swiftness of IAEA response. It is natural for a bureaucracy in a delicate political position to bend over backwards to check and re-

[81] *Supra* note 4.

[82] *Supra* note 8.

[83] Committee on International Relations, U.S. House of Representatives, "Extending the Export Administration Act," Report 94-1469, 2 September 1976.

[84] R. Rometsch, personal communication, 9 February 1977.

[85] R. Rometsch, remarks in panel discussion before the Institute of Nuclear Materials Management, 20 June 1975, reprinted in *Hearings, supra* note 18 at 1214-17.

check its figures before deciding to alarm the world (and embarrass the nuclear industry) by notifying the Security Council. A false alarm, or a true alarm that appeared false because no material was recovered or used, would be almost as embarrassing as failure to predict correctly an impending explosion. Despite the undoubted integrity and competence of the Inspector–General, bold and immediate action by any international agency is hardly to be expected, especially if a possible diversion appears only as a statistically fuzzy anomaly for which there might be many innocent—but slow to confirm—explanations.

Another fundamental limit to the effectiveness of safeguards arises, as mentioned earlier, from the statistical problems of measuring how much strategic material is or should be present. The uncertainty expected by the IAEA in assaying plutonium in a reprocessing plant is about 1 percent,[86] and improvement by more than a modest factor is unlikely for fundamental technical reasons. A covert thief is free to work within this margin of uncertainty without fear of detection by any method of accountancy. Unfortunately, 1 percent of the plutonium throughput of a large reprocessing plant is tens of bombs' worth per year. The cumulative statistical uncertainty in U.S. inventories of special nuclear material (including low-enriched uranium) is measured in tons—according to one report, about 68 metric tons[87]— of which some four to five tons, equivalent to very roughly 1000 bombs, is weapons material. It is possible that none of this material is missing; but nobody knows, or can ever know, whether it is or not, so all nuclear threats that are framed with enough technical competence to appear credible are credible.

Though, as stated earlier, a country wishing to make bombs can make its own crude reprocessing plant, commercial reprocessing plants are particularly threatening because they have large throughputs from which strategic amounts of plutonium can be embezzled without detection, because they are not safeguardable in principle,[88,89] and because—above all—they provide large amounts of weapons material whose disposition depends on sovereign national intentions. In 1975-1976, a stopgap remedy enjoyed a brief vogue: the concept of multinational regional reprocessing plants that would blunt the edge of national rivalry. After closer examination, however, particularly at the 1976 Pugwash meeting at Racine, the concept was

---

[86] *Hearings, supra* note 18 at p. 1016; similar values are reported for U.S. practice in "Oversight Hearings," *supra* note 42, Part VI, pp. 160–61.

[87] B. Newman, *supra* note 63, is broadly consistent with other reports from authoritative sources, some noted by D. Burnham, *NY Times*, 29 December 1974, p. 26; see also *ibid.*, 6 August 1976.

[88] *Supra* note 4.

[89] USHR Committee on International Relations, *supra* note 83 at 6.

quietly abandoned as politically implausible (for the same Acheson-Lilienthal reasons of national rivalry that make it hard to strengthen the NPT) and as more likely to spread the disease than to cure it: "Multinational centers for the dissemination of bomb material will not help."[90] Multinational reprocessing plants, if they could be established at all, would legitimize reprocessing and the use of plutonium, would spread reprocessing expertise to countries not now possessing it, and would make extracted plutonium—the root of the problem—widely available to embezzlers and governments alike under the same inadequate inspection safeguards that exist now. (Extracting plutonium from mixed oxide fuel is even easier than extracting it from spent fuel—yet it was precisely to prevent the latter action that multinational reprocessing was proposed!) For these reasons, the otherwise intriguing Taylor-Feiveson scheme[91] of a multinationally reprocessed thorium fuel cycle that would not use plutonium as an item of international commerce seems unlikely to succeed.

To our original question, then—whether we can have nuclear power without proliferation—the answer must be a firm no. Despite the devoted and sincere efforts of many able people for thirty years, the basic contradictions identified by the Acheson-Lilienthal report in 1946 ensure that no credible technical or political arrangements are capable in principle of reserving infallibly and perpetually to peaceful uses the enormous amounts of bomb materials that nuclear power inevitably creates. This is not to say that no steps can be taken to make abuse more difficult; rather, that over the long periods in which the bomb materials, once made, will remain potent, national rivalries, subnational instabilities and human imperfections will prove stronger than treaties and inspections.[92]

11.5. NEED WE HAVE PROLIFERATION WITHOUT NUCLEAR POWER?

"We are here," as Bernard Baruch said in 1946, "to make a choice between the quick and the dead. That is our business. . . . Let us not

---

[90] *Supra* note 8.

[91] *Supra* note 12.

[92] A red herring often introduced into such arguments is that there are many nonnuclear ways to threaten, injure, or kill large numbers of people. This is unfortunately correct. Moreover, some of these means have at least one or two—and occasionally three—of the four attributes that the nuclear threat uniquely combines, namely range in time (long-lived materials and long latency of effects), range in space, genetic transmissibility, and psychological impact. But the existence of some terrible threats is not an argument for creating more, but rather for eliminating all. Further, people insane enough to use any are unlikely to choose their threat rationally, and may find nuclear threats psychologically attractive.

deceive ourselves: We must elect World Peace or World Destruction."[93]

For twenty-three years we have tried, hopefully and often earnestly, to have nuclear power without nuclear proliferation. For the reasons the Acheson-Lilienthal panel clearly foresaw,[94] we have failed. The attempt bore in itself the seeds of its failure. "There is a melancholy lesson in the fact that today. . . we have come almost full circle in our thinking"[95]—back to the view that internationally policed national nuclear programs will inevitably spread bombs and so destroy us.

It is now time to acknowledge that "If a problem is too difficult to solve, one cannot claim that it is solved by pointing to all the efforts made to solve it."[96] Redoubled efforts to solve the same insoluble problem will be mere rearrangements of deckchairs on the Titanic. We need a new approach, a redefinition of the problem, if we are to get out of this alive.

We have only two choices. The first is to return to the Baruch Plan of 1946 and to place all potentially destructive nuclear activities—from uranium mining onward—under effective international control. This would be tantamount, in the nuclear sphere, to world government. It is an even more utopian idea now than it was in 1946. Nobody has the leverage to make it happen.

Our second choice—considered as a whole, not incrementally—is even more radical. It seems at first glance just as utopian. Jerome Frank states it thus:

> In the light of this analysis, human survival seems to require the abandonment of nuclear energy for both military and civilian purposes. The potential risks of the spread of nuclear reactors are enormous, while their only legitimate justification is as a quick source of industrial power. . . . The greatest imperative, then, is to develop alternative . . . sources as rapidly as possible. ***Unprecedented problems require unprecedented solutions. Coping with the new and unfamiliar dangers created by nuclear energy demands radically new ways of thinking and behaving.[97]

If there were no credible nonnuclear energy futures, abandoning nuclear power would indeed seem utopian. But this book is a partial survey of new insights into the energy problem and many related problems—insights that make nonnuclear futures seem both credible and realistic. The intimate connections between, indeed the indistin-

[93] *Supra* note 2.
[94] *Supra* note 1.
[95] *Supra* note 4.
[96] H. Alfvén, *Bull. Atom. Scient.* 28:5, 5-7 (May 1972).
[97] *Hearings, supra* note 18 at 82.

guishability of, the peaceful and warlike atom can make nonnuclear energy futures serve as the cornerstone of a durable peace. The rest of this chapter will consider this new approach to rebottling the nuclear genie.

First, however, we must review[98] the status of the nuclear enterprise on which President Eisenhower, with such high moral and mercantile hopes, launched us in 1953; for on the vigor, or lack of it, of that enterprise depends the practicability of reversing nuclear promotion.

During three decades of assiduous nourishment by subsidies and encouragements of all kinds, the quasi-civilian nuclear industry has never had to meet normal economic criteria. Now the chill winds of the marketplace are starting to penetrate.[99] Consider the position in the United States, with roughly half the world's nuclear business. In 1975, official forecasts called for at least twenty-six reactors to be ordered. In fact, Westinghouse sold three domestic reactors, General Electric one, and Babcock & Wilcox and Combustion Engineering none. In 1976, Babcock & Wilcox sold three reactors—not really new business—and the rest sold none. In those two years, eight reactor orders were cancelled, yielding a negative net ordering rate, while thirteen were deferred for up to four years and six were deferred indefinitely. Today no relief is in sight and the utilities are showing signs of disillusionment. The capital cost of the nuclear stations, increasing by some 20 percent/y in *constant* dollars,[100] is dauntingly high, reactor performance is probably deteriorating,[101] demand forecasts are wildly uncertain, and the technical, economic, and political problems of nuclear power loom ever larger.

Around the world the same pattern is repeated. In the U.K. the increasingly notional nuclear industry has not sold a reactor since 1970 and, according to mid-1976 statements by the Chairman and Deputy Chairman of the U.K. Atomic Energy Authority, is faced with total collapse. In West Germany there is a de facto moratorium on domestic nuclear orders pending resolution of waste problems. The French and Japanese nuclear programs have been inexorably cut back. Denmark, Norway, and Ireland have shelved plans for nuclear programs, and the Netherlands and Austria are unable to expand

---

[98] This review is largely based on, and paraphrased from, one by my colleague Walter Patterson. The events described are well known in the trade, are fully described in the trade press, and therefore are not documented here in detail.

[99] *Supra* note 8; W.C. Patterson, "The Fissile Society: Energy, Electricity and the Nuclear Option" (London: Earth Resources Research, Ltd. [40 James St. W.1], June 1977).

[100] See Chapter Two, note 3.

[101] See Chapter Five, note 21.

theirs. The Swedish Prime Minister, though later constrained by his coalition partners, was elected on a platform calling for the dismantling of Sweden's nuclear program. In December 1975, Ontario Hydro, the main customer for CANDU reactors, cut its investment plans by $5.2 billion, postponing twelve reactors and two heavy water plants (and cancelling a third outright). Further cuts are expected in Canada as in Europe.

The nuclear fuel cycle is having trouble weaning itself from the military facilities on which it has so far chiefly relied. Uncertainties in uranium supply and demand are reinforcing each other. Current enrichment capacity, especially in the U.S., is said to be already substantially committed (partly, as a comprehensive February 1977 analysis by Vincent Taylor of Pan Heuristics has shown, because of irrational but reversible ERDA policies). Yet private industry has proved unwilling to accept, and Congress to underwrite, the cost and risk of expansion. The future of multilateral European enrichment schemes is likewise uncertain: all such projects, wedding technical difficulty and market uncertainties to very long lead times and political risks, involve difficult, generally governmental, negotiations and seem increasingly precarious.

The U.K. government, at this writing, has agreed to become a £1 billion underwriter of foreign capital contributions and other loans to finance expansion of fuel cycle facilities, particularly the Windscale reprocessing plant. This gratuitous guarantee is not an idle gesture. No plant for reprocessing modern uranium dioxide fuel is in commercial operation anywhere in the world. The first such plant, opened in 1966 at West Valley, New York, consistently lost money and overexposed workers. It was shut down in 1972 for modification and expansion, but in 1976 the owner, Getty Oil, announced that the plant would not be reopened because if it were re-licensed it would have to charge $1062/kg, or forty-six times the originally contracted post-expansion price. (The $1062/kg would be only 3–4x lower if costs were spread over an expanded clientele rather than just the old one.) Apparently the state of New York will have to take responsibility for some 2.3 million liters of high level radioactive wastes at the site.

Further U.S. attempts at commercial reprocessing have failed. In 1974, General Electric had to admit that its novel Morris (Illinois) plant, after two years of attempts to commission it, did not work and probably never would. It is apparently a $64 million writeoff. Allied Chemical Corp. and partners have built a plant at Barnwell, South Carolina, next door to the Savannah River military plant; but it is years behind its planned commissioning date, may never be licensed, has already cost five times the original $50 million estimate, and will require a further $500–700 million to complete—a sum that its backers are unwilling to supply, so they are seeking a federal bail-out that they are almost certain not to get.

In the U.K., a military plutonium plant at Windscale was converted by the Atomic Energy Authority in the late 1960s to pretreat oxide fuel, but in 1973 an unexpected chemical reaction[102] expelled radio-activity and contaminated thirty-five workers. The plant has been shut down ever since for extensive reconstruction. A proposal to build a new oxide fuel reprocessing plant at the same site cannot proceed without a ministerially mandated public inquiry, and at this writing the future of the project is unclear. Indeed, the Japanese contract to use part of the proposed plant's capacity—in exchange for a major contribution to its construction cost—has not yet been signed, and neither Japan nor other countries will be able to send fuel to Windscale without U.S. government permission. This permission has recently been withheld, then granted for one shipment only; it may not be granted again, especially in view of the 9 March 1977 rule-making petition to ERDA by four environmental groups.

The proprietors of Windscale are in partnership with the French and Germans, but the small French plant at Cap la Hague is far behind schedule and has had serious labor disputes related to health and safety issues. (There has also just been an unrelated six-week strike at Windscale, during which the workers turned away supplies that would later have been needed for safety.) In West Germany, the parties desiring a reprocessing plant are unable to agree where it should be built or who should pay for it. In Belgium the small Euro-chemic plant at Mol, permanently shut down, is the subject of an

---

[102] The details of this problem are worth noting because they illustrate that reprocessing technology is far from mature. Oxide fuel of moderate burnup contains enough fission product ^{106}Ru (ruthenum–106) that it forms granules of a refractory alloy with Nb, Mo, and other rare metals. These granules are essentially insoluble in the head-end nitric acid. In the Windscale head-end accident, roughly 100 kilocuries of such granules accumulated in the bottom of a mixing vessel that was allowed to dry out between batches. The granules heated themselves to several hundred ^0C, and when the next batch of nitric acid and Butex solvent was added, they ignited and the resulting pressure forced radioactivity (chiefly ruthenium) through seals, giving one man a large lung dose (approximately 1000 rem according to one unpublished report) and others less. The Ru granule problem had not been anticipated; granules are now removed by filtration at Cap la Hague, and are to be centrifuged out in the new Windscale head-end, in order to prevent their accumulation in the plumbing. Presumably the granule problem will be considerably more difficult with fast reactor fuel than with the thermal reactor fuel reprocessed so far: partly because the burnup is about three to five times as high, and partly because the fission yield of ^{106}Ru is 0.38 percent for ^{235}U but 4.5 percent for ^{239}Pu.

acrimonious debate about who should pay for decommissioning it. And in every nuclear country, the issue of ultimate disposal of both high and low level radioactive waste remains unresolved.

Some major corporate investors are trying with difficulty to escape from nuclear commitments. It took AEG Telefunken several years to unload its half share of Kraftwerk Union, the AEG-Siemens joint venture that seems fated to continue losing money as it has done since its inception in 1969. AEG succeeded in 1976 in selling its half share to Siemens, but at unfavorable terms. Shell and Gulf, joint owners of General Atomic, withdrew from the reactor business in 1975; Shell is also unhappy with its share in the URENCO enrichment and Barnwell reprocessing ventures—both on the verge of collapse. Gulf has also had its troubles, including a major fire and two explosions in December 1972 at its Pawling, New York, plutonium fuel fabrication plant, which is permanently shut down with an undetermined amount of plutonium scattered inside and outside it. Atlantic Richfield plans to unload its management contract for the federal complex at Hanford and, like Getty, Shell, and most other oil companies, is comprehensively dismayed at the high risks and low profits of the nuclear business.

Westinghouse, the world's largest vendor of power reactors, established its dominance by sweetening some two dozen early sales with lifetime guarantees of cheap fuel. Westinghouse did not and does not have the uranium, however, and when the price unexpectedly increased fourfold by mid-1975, Westinghouse had to announce that it would renege on the fuel contracts. The liability contingent on the resulting lawsuits—some $1-2 billion—exceeds Westinghouse's book value. Meanwhile, Westinghouse is seeking treble damages from a wide range of uranium vendors (including oil companies and governments), alleging antitrust conspiracy in an extensive uranium cartel. Success in that action would presumably be reflected in the price of uranium; failure, in difficulties for Westinghouse or its fuel customers or both.

Westinghouse is the only U.S. reactor vendor that claims to be making a profit—though, suggestively, it does not publish separate accounts for its reactor business. General Electric announced in November 1976 that future sales would include far more stringent financial terms. Several U.S. vendors and architect-engineers are embroiled in disputes and lawsuits over quality control problems. The vendors, as the Vice President of General Atomic told the Atomic Industrial Forum/American Nuclear Society conference in November 1975, "have yet to make a dollar with certainty, after some twenty years of effort." Several vendors in several countries are financially so shaky that their longevity is in doubt.

   In the past three years the disappearing domestic market for reactors has prompted vigorous export drives, backed in every case by exceptionally generous government financing—in effect,[103] a device for recycling money from the Export-Import Bank or (in other countries) from taxpayers' pockets via a customer's treasury into the ailing cash flow of the domestic reactor vendor. Such indirect subsidies are now essential to even the short-term survival of the vendors. Financial terms for exports are so generous that the Tarapur reactor, for example, was exported to India virtually free. (In December 1974 France was still offering 100 percent financing of nuclear exports at interest rates as low as 6.3 percent/y, repayable over fifteen years.) Many exporters have, in effect, paid their customers to haul the reactors away: early contracts commonly involved loss-leader turnkey contracts, cheap fuel guarantees, no interest loans, long grace periods, and other considerations. German, U.S., and Canadian reactor sales to Argentina, India, and Pakistan, respectively, all incurred heavy financial losses for the vendors. Further, the Canadian Auditor General reported in November 1976 that some $10 million might have been used as bribes to facilitate CANDU sales to Argentina and South Korea, the only two "open market" export orders Canada has ever won. But some customers are not so lucky: it is said that French vendors are selling reactors to Iran and South Africa at approximately two and three times, respectively, the price they charge Electricité de France.

   If the gloomy picture painted here—as in the nuclear trade press— is accurate, then for the nuclear industry to survive, let alone expand at least twentyfold over the next few decades, requires vigorous and prolonged action by many governments. Vigorous action is more than just tolerance; it costs a great deal of money and requires both public and private political support. Such support is as necessary in the most outwardly confident European nuclear program as in the obviously troubled U.S. program. The former look healthy from across the Atlantic only because of distance, not familiarity: likewise, to a French or German observer who only reads ERDA press releases, the U.S. program might also appear, erroneously, to be brushing off a minority fringe protest and going from strength to strength. The French nuclear program is the archetype of single-mindedness as seen from abroad, but from within it is clearly the subject of serious disputes both inside and outside the central government. In 1975, Irvin Bupp was even told[104] at the highest levels of

---

[103] W.C. Patterson, "Exporting Armageddon," *New Statesman* 92:264-66 (27 August 1976). For one effect, see M. Liu, *Washington Post*, 27 February 1977, p. A21.
[104] See Chapter Two, note 33, and Chapter Eight, note 11.

the French nuclear program that there was perhaps a one in three chance that most of the planned reactors would actually be built—and an equal chance that the program would collapse. (Since then the fiscal and political constraints on the program have tightened markedly: according to a well placed French official, four of the six politically "safe" sites, by early 1977, were already in trouble.)

The key to reversing nuclear promotion lies, then, in the lack of a domestic political base for the strong and sustained government actions needed in each country to shield proposed nuclear programs from political and economic realities.[105] Not only is this broad base of enthusiastic support lacking; there is outright and, in many cases, formidable opposition that interacts across national boundaries. Chapter 2.8 argued that the attitude of the United States in particular will be crucial to those disputes: whatever the U.S. does will help one side of the nuclear debate and hurt the other. My own political judgment is that the U.S. government is exceedingly unlikely to bail out its own nuclear industry on the required scale; and that an active renunciation of nuclear power within the next few years is politically plausible. Indeed, I am confident that because of the political and economic forces already set in motion, nuclear power is politically and economically dead in North America.[106] The key question—one of which I am only slightly less confident—is whether the demise of the program will be put in a constructive context, as Chapter 2.8 suggests, closely linked with soft energy policies, nonproliferation, and strategic arms reduction.

It is worth reviewing the U.S. actions that should, if Chapter 2.8 is correct, provide the new approach we need to the whole nuclear problem:

1. phase out the U.S. nuclear power program and U.S. support of foreign nuclear power programs;
2. divert those resources to implementing and promoting soft energy paths, freely and unconditionally helping any other interested country to do the same;

---

[105] See Chapter Two, note 31, for the possibly anomalous case of the USSR. The dissent within the Soviet scientific community has reached such a pitch that the Communist Party has felt compelled to issue a strong statement endorsing the nuclear program. The effect of such formal commitments on the program (as opposed to the dissenters) is largely academic, however, since the Soviet Union is suffering from an even worse capital shortage than Western countries.

[106] Dead in the sense of a brontosaurus that has had its spinal cord cut, but is so big and has so many ganglia near the tail that it can keep thrashing around for years not knowing it's dead yet.

3. treat control of nuclear power, nonproliferation, and strategic arms reduction as a single integrated problem, not riven by an artificial civilian-military distinction: in short, adopt the Acheson-Lilienthal view[107] that it is futile to expect national nuclear programs, however policed, to stay peaceful, and try therefore, gradually but relentlessly, to eliminate all of them together.

The object of these actions, taken simultaneously, unilaterally (at least at first), and uncompromisingly, would be to foster an international psychological climate of denuclearization in which it comes to be socially unacceptable—perceived as a mark of national immaturity—to have or seek either reactors or bombs. But this could work only if the three steps are taken together and linked together: *their linkage has a psychological synergism that is essential to their success.* Of course success may be slower in reducing military than quasi-civilian nuclear activity, owing to the lesser interests at stake and the greater openness to political control in the latter case, but both must be pursued together and thought of together if either is to succeed.

Nor should the United States be bashful about leadership in this unusual effort. As Herbert York remarks: "The present mess—if we may call it that—was created by the unilateral initiatives of the United States. It should not surprise anybody if, in the unravelling of this mess, it requires unilateral initiatives by the United States."[108] Indeed, a propitious beginning would be for the U.S. to state clearly that the Atoms for Peace program begun in 1954, though well meant, was a serious mistake that the U.S. is now willing to go to a great deal of trouble to correct.

The unique ability of the U.S. to effect denuclearization unilaterally, however, complements, rather than detracting from, the ability of other countries to make signal contributions. The Swedish nuclear debate is still reverberating around the world. Many labor unions in Australia, concurring with the Fox Commission that the nuclear industry is "unintentionally contributing to an increased risk of nuclear war," are opposing uranium mining, particularly at the major deposits in the Northern Territory; this courageous subordination of private to public interests is rightly beginning to attract worldwide attention, the more so because anything affecting 20–25 percent of potential world uranium supplies is more than a mere gesture. Similar issues, overlain by colonialism, are arising in Namibia, and uranium mining and export are becoming an issue in Canada. The nuclear debates in Denmark, the Netherlands, France, West Germany,

[107] *Supra* note 1.
[108] *Hearings, supra* note 18 at 41.

Switzerland, Britain, New Zealand, Japan, and elsewhere are reinforcing each other with new technical, political, and moral authority. Even the smallest country can add disproportionate political weight to the world debate. Thus nuclear restrictions everywhere help each other and, ultimately, bring pressure to bear on critical decisions in the United States, a country often sensitive to world opinion and example.

Some elements of the denuclearization program considered here for the U.S. have been proposed before. There is, of course, a vast literature on ways to inhibit military nuclear competition—various combinations, for example, of increasingly significant restrictions on testing, deploying, and using bombs, each unilaterally introduced on schedule if the previous step meets with an encouraging substantive response from other nuclear weapons states. But distinguished commentators have also suggested, generally piecemeal, some dramatic steps toward civilian denuclearization. For example, in January 1976 David Lilienthal, first chairman of the USAEC and author of the Acheson-Lilienthal report of thirty years earlier, told the Ribicoff Committee of the U.S. Senate that:

> . . . in spite of these years of effort, the tragic fact is that the atomic arms race is today proceeding at a more furious and a more insane pace than ever. Proliferation of capabilities to produce nuclear weapons of mass destruction is reaching terrifying proportions. And now, the prospect of the reprocessing or recycling of nuclear wastes to produce weapons material from scores of atomic power plants is close upon us. . . . We have to decide now what we can do, now, within our own capabilities, to prevent a very bad situation from becoming a disastrous and irreversible one in which the whole doctrine of atomic weapon deterrence which underlies our entire military policy falls.
>
> I therefore propose as a private citizen that this committee. . . call upon the Congress and the President to order a complete stop to the export of all nuclear devices and all nuclear material, that it be done now, and done unilaterally. Further, unilaterally, the United States should without delay proceed by lawful means to revoke existing American licenses and put an end to the future or pending licensing to foreign firms and government of American know-how and facilities . . . .
>
> This action taken alone will not solve the whole problem of proliferation. But it will put an end to our major part in it. For the fact is that we, the United States, our public agencies and our private manufacturers, have been and are the world's major proliferators.
>
> Admittedly, an immediate embargo on the export of nuclear technology[,] materials, and plants. . .would be drastic action, so drastic as to raise the question of America's moral position to take such a course. . . .
>
> Certainly, we wish we could believe that the policies and actions we think necessary to protect ourselves and the world might meet with the

complete and ready approval of other countries. Whether they do or not, the history of American participation in nuclear development in the past 30 years confirms that our moral position is a strong one, and it entitles us to proceed even if the approval and emulation of others are not immediately forthcoming. . . .

[W]e have earned the right, we have the responsibility and we have the leadership to put our own house in order . . . and to decide what we should or should not export to others.

. . .If we adopt. . . [a] defensive, apologetic view of America's posture respecting the further spread of atomic weapons, you can be sure that once more we will be taken advantage of, because of our national inclination to want to be loved in foreign parts, even at the cost of being the atomic patsy of the world—which is what we have become.

I suggest that we should not be overly impressed by the morally indefensible doctrine that if our manufacturers and vendors do not continue to supply these potentially deadly materials and this technology, the manufacturers of other countries will do so.***

[W]e can provide for. . . a moratorium: a ringing declaration by the Congress against exporting a single additional gram of plutonium or enriched uranium or a single additional nuclear reactor of any kind by anyone from this country to any other country, until and unless genuinely effective internationalization is a fact.***

No one is wise enough to foretell what would happen if now the people of the United States put a flat and complete embargo on the export of nuclear technology, facilities and materials. But what is our alternative?

Our experience in international negotiations on the whole has so far been a dismal one. If we now show that we mean business, it may very well improve.[109]

Herbert Scoville, a scientist with distinguished service in the U.S. military and intelligence community, has stated his concern that nuclear weapons

. . . will soon fall into many hands in many corners of the world—into the hands of unstable national governments, aggressive military cliques or irresponsible terrorist groups, with incalculable consequences for us all. This danger is the direct result of the uncontrolled growth of the nuclear power industry which is making widely available the materials needed for such weapons. The peoples of the world must recognize the danger of what is going on and act to protect this and future generations.

***I believe we must create an international atmosphere where the possession of nuclear weapons is a cause for embarrassment and shame—rather than for power and prestige.[110]

[109] *Ibid.*, p. 10.
[110] G. Kistiakowsky et al., "The Threat of Nuclear War," Granada TV program, 29 March 1976 (transcript reprinted May 1976 by Granada TV, Manchester, U.K.).

George Kistiakowsky, who headed the Explosives Division of the Manhattan Project and was President Eisenhower's science advisor, then added:

> [I]f no action is taken, the spread of nuclear technology will soon advance beyond the ability of human societies to control it . . . .
>
> Unfortunately, our governments are beset by many problems which they see as more immediate. They are naturally also under pressure of their commercial interests. In their eagerness to see trade expand, governments are not willing to override their short-term interests to gain long-term nuclear peace for the world. A recent agreement among nuclear seller countries, allegedly aimed at preventing proliferation, provided only token controls.
>
> The present gradual slide of humanity into nuclear war can be stopped only by the voice of the people. And it can be stopped, for nuclear installations take many years to build, and there is still time to act. A broad public movement must tell the governments of the seller countries that present policies of foreign trade in nuclear technology must change.
>
> First, there should be no sales of fuel or processing plants.
>
> Second, the spent nuclear fuel, containing plutonium that could be used for bombs, must be put into storage facilities under international control.
>
> Unless we take these first steps we will have lost the chance of avoiding a nuclear holocaust.[111]

The Committee on International Relations of the U.S. House of Representatives addressed in September 1976 the role of U.S. initiatives in nuclear export policy. The committee concluded:

> The United States, which has long enjoyed the lion's share of the world nuclear export market, quickly arouses the commercial suspicions of other supplier nations when it attempts to convince those nations to control their own exports. Only when we clearly enunciate U.S. nuclear export policy will these commercial suspicions diminish. It is difficult, for example, to convince France to refrain from exporting reprocessing facilities as long as U.S. intentions concerning its future export of reprocessing plants and its willingness to permit other nations to reprocess U.S.-supplied fuel remains [*sic*] undefined.
>
> Some argue that more restrictive U.S. export policies will only result in our being driven out of the nuclear market, thereby eroding further our influence over both supplier and recipient states. This argument is too narrowly commercial. It assumes that the supplier states—most of whom are our military allies—share no common interest with us in reducing the ease with which other states can acquire threatening nuclear weapons capabilities. This is clearly not the case. Moreover, this argument also

[111] *Ibid.*

assumes that there is no longer a place for leadership by persuasion and examples. As the nation which stimulated today's broad distribution of critical nuclear knowledge and technology, we have a special obligation to provide enlightened leadership in this area.[112]

Canada, too, has set an important precedent by withdrawing nuclear cooperation from India and Pakistan[113] and by announcing on 22 December 1976 that she will not export any nuclear reactors or materials to countries that have not ratified the Non-Proliferation Treaty or accepted "full scope" safeguards. On that occasion Prime Minister Trudeau called on other nuclear exporters "to review their own export policies, not in the light of commercial gain but in the interests of maintaining a safe and secure world." The full impact of this precedent, like that set by the Australian unions, is yet to be felt; but Canada, significantly, caught even the EEC in her net.

As Lilienthal states,[114] the U.S., even more than other nations, is peculiarly well placed to influence world nuclear affairs because her leverage is far more than just political and institutional. The U.S. now controls, and will continue for about another decade to control, most of the non-Communist world's supply of enriched fuel, including nearly all fuel for the EEC. Probably every light water reactor program and manufacturer in the world depends in some way on an intricate and little known network of U.S. licensing and technical support, all of which is probably subject to export controls that have not yet been much exercised. As Herbert York points out:

> We do dominate the field. Something like 50 percent of all the activity in the nuclear field is American activity, no matter what aspect of nuclear activity one looks at.
>
> So we have a very large handle on it. It is not as if we were sitting in Belgium discussing some kind of moratorium. We here have real influence. . . .
>
> There is also the way . . . the technology is interconnected in ways that are hard to anticipate, so an American moratorium would probably very much influence not only what the French would do, but what they could do.
>
> If I may take an example. . . , Soviet reactors in Finland. . . are being built by the Soviet Government as a principal contractor. But the backup safety system is being supplied by Westinghouse Electric. The Finns regard

---

[112] *Supra* note 83.

[113] The oft-cited "obligation" under Article IV of the Non-Proliferation Treaty to proliferate civilian nuclear technology is not an obstacle to withholding it: see *Hearings, supra* note 18 at 144, 1391, and 1402; Lowrance, *supra* note 14 at 159; and reread the text of Articles I and IV of the NPT.

[114] *Supra* note 109.

this as an essential element in the building of a reactor. Thus, doing without the Americans is difficult even in the case of a Soviet-supplied reactor. They need us as subcontractors.[115]

The pervasive web of licensing agreements—just starting to be mapped—produces many surprises of this kind. Even the most outwardly independent European reactor program today requires U.S.-licensed processes for manufacture or controlling water chemistry, U.S.-made instruments, U.S.-designed hardware of many kinds. Even the supposedly independent Canadian CANDU program exists only on U.S. sufferance: some 20–30 percent of the hardware is made in the U.S., including Zircaloy calandria tubes that require a U.S. proprietary process. Most large CANDUs also require U.S.-enriched uranium for booster rods—weapons grade uranium whose export was recently held up owing to U.S. concern about inadequate Canadian physical security.

In short, U.S. influence even in the purely technical sphere of designs, equipment, processes, and fuel cycle services is enormous. The bare possibility of its exercise has been enough to reverse French and German positions on the lucrative export of fuel cycle facilities[116]—and it now appears unlikely, despite official statements to the contrary, that these sensitive technologies will in fact be delivered to Pakistan and Brazil as proposed.

If the U.S. is to cease exporting all nuclear facilities, whether through more stringent export licensing or through removing the essential Export-Import Bank financing, the question must naturally arise how the U.S. can justify using reactors herself. It is a good question to which there is no answer. The U.S. cannot say that she, unlike other countries, is responsible enough to use nuclear power wisely, for other countries can legitimately point to major examples of U.S. irresponsibility from Hiroshima onwards. Symmetry—the principle that what one says is bad for others must also be bad for oneself—requires that the U.S. also stop building reactors at home and begin with others to devise an orderly plan for the terminal phase of the nuclear economy. This raises many nice technical puzzles: for example, whether it might even be worth building some novel type of low power density reactor especially to burn up as much as possible of transuranic stocks (at first civilian and later, as soon as possible, military). The technical community, liberally aided

---

[115] *Supra* note 108.
[116] W.D. Metz, *Science 195*:32 (1977).

by wider constituencies, must start to address such questions. Nobody has ever done so before.

Getting out of the nuclear business also implies shifting simultaneously to a sound nonnuclear energy policy—the burden of this book—and developing ways to recycle the nuclear industry.[117] That does not mean bailing out the corporations involved: they have taken a business risk in full knowledge of the economic consequences of failure (from whatever cause), and many are already said to be shifting their emphasis back to more traditional pursuits such as nonnuclear boiler making as their nuclear orders fall off. But nuclear vendors can make other things too. So long as they retain their technical and managerial skills, they will always make a profit—probably a larger profit in pursuits that strain those skills less harshly than nuclear power does. The vendors' skills and productive capacities will be needed in energy, transport, and the like; but the vendors can largely be left to find new markets for themselves, as they have their own contingency plans and are big enough to look after their own interests.

The 60,000-odd U.S. technologists who, in good faith and with the best intentions, took up a technical challenge only to be told its solution was no longer wanted do deserve help during the transition. The same skills that make reactors are desperately needed elsewhere—and can be applied elsewhere. Most nuclear technologists are not highly specialized, but are rather good electrical, mechanical, chemical, or other kinds of engineers with enough breadth and adaptability to have gone into the nuclear field. They are also adaptable enough to change again, just as the engineers who put men on the moon are now engineering simpler earthbound devices. The highly specialized minority of nuclear technologists—radiochemists, reactor physicists, and so on will be well occupied minimizing the residual hazard of the nuclear industry (a career, as Walter Patterson remarks, with a future—a long future). Anyone who still has nothing to do should be retrained at public expense: nobody with the skills and drive that the nuclear industry has displayed should lack a chance to put them to better use. The problem we face here is just like that of converting aerospace and defense industries. It is a problem we shall have for decades, and we should start getting used to it.

Utilities that have sunk billions of dollars into reactors—and are therefore having to raise astronomical sums from bondholders, rate-

---

[117]The following material is adapted from the author's contribution to D.R. Brower, ed., *Politics As If Survival Mattered: A New-Era Testament for America* (San Francisco: Friends of the Earth, 1977). Some needed actions are described there in greater detail.

payers, taxpayers, and everyone who pays interest on the money that the utilities' investments make scarcer—presumably bought reactors with full knowledge of the risks (as they often remind us). They should therefore be prepared for substantial losses if they invested imprudently—or if they failed to share their knowledge properly with their stockholders. But the alternative to those losses is far larger losses, and an utterly unsupportable cash flow in the long run, if more money is instead poured down the fission rathole to raise the nuclear share of the U.S. electrical capacity above its present 8 percent. Nuclear-powered utilities would generally lose less money, and have to raise prices less, if they rapidly wrote off their relatively small stock of nuclear reactors, complete with their economic disappointments—starting with the reactors that are nearest to cities and most dangerous—than if they persisted in grandiose plans for nuclear expansion. Their money will be much better spent in financing cogeneration, district heating, conservation, and soft technologies.

Nor are past nuclear investments wholly lost. The nuclear part of a power station represents only about a tenth of its capital cost, and, as Sweden has shown at Marviken, the rest can generally be salvaged by conversion (if it is needed at all) to fossil fuel—preferably a coal-fired fluidized bed. Even more investment and jobs can be salvaged by stopping and, in many cases, altering nuclear stations already under construction. So far, fortunately, little capital has been sunk into fuel cycle facilities to supplement or replace those left over from military programs; so here too it is an opportune moment to cut our losses rather than increase them further.

Phasing out nuclear power should make our electricity cost not more but less. Nuclear electricity from new reactors, even in the narrowest sense of traditional economics, will cost substantially more than from other sources (ranging from conventional coal-fired power stations to the much cheaper cogeneration systems) barring the lopsided and unrealistic assumptions beloved by vested interests.[118] (This comparison ignores the social costs on both sides.) Even today the economic advantage of nuclear power, given honest accounting, is speculative and at best minute.[119] And nuclear power, or any other kind of electricity, is vastly more expensive than increased end–use efficiency.

[118] For realistic or even conservative assessments, see, e.g., Chapter Three, notes 19–20; Chapter Five, note 21; Chapter Six, note 25 (and the book *Light Water* by Bupp and Derian, New York: Basic Books, 1977); and "The Nuclear Power Alternative," Special Report 1975-A (New York: Investor Responsibility Research Center, January 1975).

[119] For example, the Nineteenth Steam Station Survey (see Chapter Five, note 20) showed U.S. nuclear electricity from typical recently completed plants costing about a third more than coal-fired electricity. In view of Chapter Six, note 39 it is not clear why any of the reactors were built.

An even more urgent step than phasing out existing reactors will be *irrevocably* forswearing the plutonium economy. Again, the U.S. can hardly persuade or expect others not to breed, reprocess, and recycle plutonium if she appears to be retaining those options. Accordingly, the U.S. fast breeder program and proposed licensing of reprocessing and recycling should be *unambiguously terminated*, and this action should be explicitly linked to global denuclearization.

Foregoing the proposed next steps in the plutonium economy— and even the breeder program itself[120]—will have economic advantages. Reprocessing offers no prospect of economic gain for many years.[121] Reprocessing introduces significant health and safety risks,

[120] See, e.g., "Oversight Hearings," *supra* note 42, pt. II; T.B. Cochran, *The Liquid Metal Fast Breeder Reactor* (Baltimore: Resources for the Future/Johns Hopkins University Press, 1974); Joint Economic Committee, U.S. Congress, *Fast Breeder Reactor Program*, Hearings 30 April and 8 May 1975 (Washington, D.C.: USGPO 64-603, 1976), especially T.B. Cochran et al., "Bypassing the Breeder," at pp. 470-543 Comptroller General of the U.S., "The Liquid Metal Fast Breeder Reactor: Promises and Uncertainties," OSP-76-1, U.S. General Accounting Office, 31 July 1975; B.G. Chow, "The Liquid Metal Fast Breeder Reactor: An Economic Analysis" (Washington, D.C.: American Enterprise Institute for Public Policy Research, December 1975); I.C. Bupp & J.-C. Derian, *Techn. Rev.* 76:8, 26-36 (July/August 1974); E.W. Carpenter et al., "Sodium-cooled fast reactors: an electricity utility's perspective," *Proc. Intl. Conf. Fast React. Pwr. Stns.*, British Nuclear Energy Society (London), 14 March 1974, p. 631; T.B. Cochran, "An Alternative LMFBR Program" (Washington, D.C.: Natural Resources Defense Council, 28 February 1977);M. Sharefkin (Resources for the Future, Inc.), "The Fast Breeder Reactor Decision: An Analysis of Limits and the Limits of Analysis," Joint Economic Committee, U.S. Congress, 19 April 1976; USEPA, "Comments on Proposed Final Environmental State ment, LMFBR Program," Chapter V ("Cost Benefit"), D-AEC-00106-00, April 1974, reprinted in USAEC, WASH-1535, vol. 7, pp. VII. 53-32 through VII. 53-80 (December 1974) and summarized by R. Gillette, *Science 184*:877 (1974); and a wide range of other studies whose references are available from T.B. Cochran, Natural Resources Defense Council (917 15th St. NW, Washington, D.C. 20005).
[121] See, e.g., Wohlstetter *supra* note 8; V. Taylor, prepared statement in California Energy Resources and Development Commission testimony to USNRC GESMO hearings, 4 March 1977; W.D. Metz, *Science 196*:43-45 (1977); S.M. Keeney et al., *Nuclear Power Issues and Choices* (Cambridge, Massachusetts: Ballinger for the Nuclear Energy Study Group of the Ford Foundation, 1977); USHR Committee on International Relations, *supra* note 83; Kistiakowsky, *supra* note 110; M. Resnikoff, in "Oversight Hearings," *supra* note 42, pt. I, pp. 847-97; B. Wolfe & R. Lambert, "The Back-End of the Fuel Cycle," AIF Fuel Cycle Conference, 20 March 1975; ERDA, "Nuclear Fuel Cycle," ERDA-33 (1975) at ix; and, for the ^{236}U penalty, E.W. Colglazier & R.K. Weatherwax, "The U236 Penalty for Recycled Uranium," COO 2220181, Princeton University, September 1976; International Technology Project, *Non-Proliferation and Nuclear Waste Management*, study for ACDA (contract AC6AC725) (Berkeley, California: Institute of International Studies, University of California, 1977), Appendix E. Those studies that claim an economic benefit for reprocessing admit it is very small—of the order of 1 percent on busbar cost, less on consumer price—but appear to underestimate the marginal cost of reprocessing (several hundred $/kg) and perhaps to overestimate the forward price of uranium. Moreover, recent results of the USNRC

both public and occupational. Plutonium recycle would save some ^{235}U, but when calculated for a growing reactor program rather than for one reactor alone, the uranium saving would only stretch the useful life of thermal reactors by a few years.[122] Moreover, reprocessing probably makes waste management harder rather than easier, for it *increases* the total amount of waste which is transuranic contaminated and must therefore be disposed of as if it were high level waste. That is, the total volume of such waste—which is what matters for waste disposal, both logistically and economically—is about twice as great after reprocessing as before, especially if subsequent oxide conversion and fabrication are considered.[123]

Moreover, the shifts of composition of the spent fuel during reprocessing are also troublesome.[124] The thermal power is almost entirely concentrated in one stream whose ultimate volume is an order of magnitude smaller than that of the spent fuel; the resulting high density of thermal power is vexatious. The plutonium is split three ways: extracted plutonium, which does not go away (it or the especially actinide-rich wastes produced by fissioning it must still be disposed of eventually), residual plutonium in high level wastes, and extensive low level wastes. The amount of plutonium in the latter two categories is somewhat less than one order of magnitude smaller than in the spent fuel (not two orders, owing to decay of higher actinides); but that quantity is roughly half in chemically and physically diverse low level wastes—which are harder than spent fuel to encapsulate and isolate, have larger surface area and often a lower chemical binding energy, and are probably more prone to leaching, weathering, bacterial action, etc. (Even the long-term stability of vitrified high level wastes is unproven and controversial.) Thus any quantitative improvement in actinide problems by reprocessing spent fuel is likely to be at least offset by a qualitative worsening.

The consequence of this argument is that the option of *never* reprocessing must be kept open by suitable storage methods, preferably at sites convertible to ultimate disposal sites. Any method of

---

security force study suggest that the costs of such a force will roughly equal the benefits claimed for Pu recycle by its proponents. No public review of reprocessing costs and benefits appears to be available outside the U.S., though in Britain the Chief Executive and the Chairman of British Nuclear Fuels Ltd. stated publicly (the latter reprinted in *Atom*, January 1976, p. 9) that reprocessing oxide fuel in the U.K. is not currently economic. For a British view of the issues see *Nature 264*:691–69 (1976).

[122] International Technology Project, note 121 *supra*, Appendix D.

[123] G.I. Rochlin, testimony to GESMO Hearing Board, USNRC (Docket No. RM–50–5), 4 March 1977; and Chapter 4, "Is Reprocessing Required for the LWR Fuel Cycle?," in International Technology Project, note 121 *supra*.

[124] The arguments in this paragraph are documented by Rochlin, *ibid*.

waste disposal that is satisfactory for high level reprocessing wastes, in whatever physical form, will also be satisfactory for appropriately packaged spent fuel. Thus there are no persuasive economic, resource, or waste management arguments for reprocessing over at least the next decade or two (if not longer), and there are transcendent proliferation arguments for permanently abandoning reprocessing.

Foregoing any possibility of a plutonium economy—an unambiguous renunciation that will have insignificant economic and social costs —is an essential beginning, and there are cogent, detailed arguments for doing it now. But important though those arguments are, it is even more important not to lose sight of the wholeness of the nuclear problem. We need to see our proliferation, disarmament, development, energy, and social problems in their full complexity and interrelatedness. And we need to add a more acutely political and global perspective to our economic and technical ones. Thus, while unilateral action to discourage the plutonium economy would not absolutely prevent non-NPT countries such as India[125] from exporting their own sensitive facilities,

> . . . the likelihood of their doing so must depend greatly on other actions of the present nuclear exporter states. If the nuclear weapons powers were to end all nuclear weapons tests and show genuine progress in limiting their nuclear armories, and if technical aid were forthcoming in really substantial amounts to help discover new energy sources . . . [including] development of solar, geothermal, wave and other renewable energy sources, then the moral and political pressure against non-NPT states pre-

---

[125] India is probably the best example of a potential pocket of metastatis. It remains to be seen how thoroughly her nuclear industry could be self-sufficient in view of the difficulties at, for example, Tarapur (see *Hearings, supra* note 18 at 655–86). Even more doubtful is her ability to withstand political and economic pressures, especially from within, as opinion even inside the nuclear program shifts toward the view that electrification is irrelevant to the needs of most of India's people. (Electricity in India, as in most developing countries, goes typically 80 percent to urban industry, 10 percent to urban homes, and 10 percent to rural villages—where 80 percent of Indians live. Even that 10 percent goes typically to only 10 percent of the village homes, i.e., to the few relatively rich villagers who were not the object of the exercise.) Of course, India already has bombs and will want to keep them as long as others do. But the question is whether India could and would both maintain the domestic base (political, economic, and technical) needed to become an independent nuclear exporter and find qualified client states seeking her exports, all within the political and psychological climate of denuclearization. This seems implausible. And even Indian technical ability, which is substantial, would be hard pressed to maintain a substantial domestic program for more than a decade or two in the face of an effective quarantine such as Canada has begun. Recent demolition (see note 6, *supra*) of the grotesque notion that nuclear power is a necessary, appropriate, or economically attractive energy source for developing countries makes such secondary metastasis seem even less likely.

pared to proliferate the plutonium economy would be so great as to make it unlikely that they would wish to pay the price. As to the Communist bloc countries, the Soviet Union, especially, has since the days of the first Chinese tests . . . shown great caution regarding nuclear proliferation, and judging from its participation in the NSG [Nuclear Suppliers' Group], will be among those nuclear exporter states most cautious about the spread of the more weapons-sensitive parts of the nuclear fuel cycle.[126]

Until now we have treated nuclear power and nuclear armaments as unrelated, and so sown dragons' teeth to the four winds. We have acted out on a world scale Paul Ylvasaker's definition of a region as "an area safely larger than the one whose problems we last failed to solve." By treating the civil and military atoms as distinguishable, and our own possession of bombs as patriotic but others' as irresponsible, we have woven a web of hypocrisy and doublethink that is as inimical to arms control as to nonproliferation. The former, which is the greater and harder problem, cannot be addressed without a new beginning on the latter—without a fresh start at attacking the tensions and inequities that are at the root of the East-West and North-South arms race. But the same promotional skill that spread reactors around the world can now proliferate alternatives to those reactors and so place prohibitive political obstacles in the way of the nuclear future. The same ingenuity and good will that managed, despite the inherent illogicalities and contradictions of proliferative nonproliferation, to obtain the small measure of international agreement that we have today on nuclear issues can now, freed from commercial imperatives, find ways to keep ominous trend from becoming fatal destiny. Perhaps the Acheson-Lilienthal message will yet be heard in a new setting:

[W]e have outlined the course of our thinking in an endeavor to find a solution to the problems thrust upon the nations of the world by the development of the atomic bomb—the problem of how to obtain security against atomic warfare, and relief from the terrible fear which can do so much to engender the very thing feared.

As a result of our thinking and discussions we have concluded that it would be unrealistic to place reliance on a simple agreement among nations to outlaw the use of atomic weapons in war. We have concluded that an attempt to give body to such a system of agreements through international inspection holds no promise of adequate security.

And so we have turned from mere policing and inspection by an international authority to a program of affirmative action. . . . This plan we believe holds hope for the solution of the problem of the atomic bomb. We

[126] *Supra* note 7.

are even sustained by the hope that it may contain seeds which will in time grow into that cooperation between nations which may bring an end to all war.

The program we propose will undoubtedly arouse skepticism when it is first considered. It did among us, but thought and discussion have converted us.

It may seem too idealistic. It seems time we endeavor to bring some of our expressed ideals into being.

It may seem too radical, too advanced, too much beyond human experience. All these terms apply with peculiar fitness to the atomic bomb.

In considering the plan, as inevitable doubts arise as to its acceptability, one should ask oneself "What are the alternatives?" We have, and we find no tolerable answer.[127]

Need we have proliferation without nuclear power? Not if we do it right. The methodical collapse of nuclear power under its own weight now offers us, briefly, the chance to start afresh; the chance to get the world speedily and permanently out of the nuclear business; the chance to go back and deal more wisely with the consequences of the first Promethean visit before we are incinerated by the second. But the political opportunity we now enjoy will soon pass by and will never come again. We must stop passing the buck before our clients start passing the bombs.

Not only the chance to choose right but the chance to choose at all is unique and fleeting. Putting the nuclear genie back in the bottle before it puts us in our caskets is a prerequisite for everything else, the most urgent issue of this or any age. So Chapter Two argued that we must choose between the soft and hard energy paths "before failure to stop nuclear proliferation has foreclosed both." And just as the nuclear choice lies near the center of the energy problem, so in the choice of energy paths lies the key to denuclearization. only the soft path gives us the alternatives and the political leverage to make it work. The hard path would instead sow thousands of tons of plutonium to the corners of the earth; and a durable peace cannot be built out of an even more durable bomb material. The contrast between plutonium and sunbeams could not be more complete. "It says much for the ambivalence of the atom," said Cousteau[128], "that we ever had to have an Atoms for Peace program. How gratuitous it would be to suggest a Sunbeams for Peace program! Mankind has probably not used sunbeams for war since the Battle of Syracuse."

The gifted technologists who first put the primal nuclear fire into our untutored hands now wish to elaborate its details, turn it to new

[127] *Supra* note 1.
[128] J.-Y. Cousteau, opening remarks, *supra* note 10.

tasks, thrust it—as saviors—on the sober and the reckless alike. Why do people with the most reason to know of human folly and ruthlessness show such generous faith in human wisdom and self-control? Perhaps, understandably, they are bewitched by their own glittering achievements. Perhaps, fascinated by their unravelling of the innermost secrets of the physical world, they are less mindful of the unchanging but more opaque truths of life, mind, and spirit. For their discoveries have transformed only one narrow, abiotic aspect of the world. Despite cataclysmic advances in mankind's ability to wrest useful work from the insides of atoms—or to ignite miniature suns halfway around the world—

> . . . the basic facts of life on earth remain the same. The way the world works and has worked for a few billion years, the way people behave, the way chemical energies are driven by sunlight to shape living systems ever more stable and intricate—all are still the same as they ever were. . . .[129]

In that continuity lies the key to our survival: working with, not trying to smother and replace, the life force that has brought us to this place.

> [S]unlight, in its many guises, is the force that has shaped and driven the miraculous living fabric of this planet for billions of years. It embodies the best engineering, the widest safety margins, and the greatest design experience we know. It provides amply for our needs, yet limits our greed. . . .
> It is safe, eternal, universal, and free. It falls justly and equitably on South and North, East and West. It increases autonomy, fosters diversity, and does not hurt the balance of payments. Its quality is constant and very high.[130]

Sunlight leaves an earth unravished, husbanded, renewed. It leaves a people unmutated, convivial, even illuminated. Above all, it respects the limits that are always with us on a little planet: the delicate fragility of life, the imperfection of human societies, and the frailty of the human design. We can still choose to live lightly, to live with light, and so choose life itself—by capturing the Hope left waiting at the bottom of Pandora's box.

[129] *Supra* note 128.
[130] *Ibid.*; T.B. Taylor, *Skeptic*, March/April 1977, p. 43.

✳

Afterword

The psychological climate surrounding nuclear power is changing rapidly. Public and disinterested professional support for the technology—as several analysts have suggested, and as a Canadian Nuclear Association poll in 1976 confirmed—is confined increasingly to the uninformed. This shift results no less from wider knowledge of the technology's hazards and uncertainties than from wider knowledge of alternatives to it. Here the mood has shifted most dramatically. In the early 1970s a relatively small group of enthusiasts was urging that conservation and solar technologies be studied to see if they contained major options worth serious examination; but now it is the nuclear lobby that is pleading for heroic measures to keep its rapidly fading option alive in case it might someday be needed. This winnowing of arguments on their merits can only continue to shift the balance as public knowledge and discussion of energy policy expand.

Meanwhile, the very nature of discourse about energy policy, the type of questions asked, the breadth of answers expected, has broadened beyond recognition. Energy policy now embraces the macroeconomic, the geopolitical, the anthropological, and even the moral sphere, and the social details of energy use are widely considered to be much more interesting and important than the technical options for energy supply. The effort made by this book and its predecessors to bring some modest synthesis to the enormous ferment and flux of energy thinking around the world seems by luck to have come to a focus at the right moment. In a sort of snapshot, therefore, this Afterword will try to place soft energy path research within the context of ongoing criticism, refinement, and extension in the U.S. and abroad.

The paper in *Foreign Affairs* (October 1976) that was reworked into Chapter Two of this book has been widely circulated and discussed. Its concepts were the subject of seminars at many universities, government departments, international agencies, and energy (mainly nuclear) research centers during 1975–1977. The paper itself was the subject of a debate between the author and Professor E. Linn Draper (Nuclear Reactor Laboratory, University of Texas at Austin) before a special joint hearing of the Small Business and Interior Committees of the U.S. Senate on 9 December 1976. The record of that hearing, in press for USGPO publication in spring 1977, also includes comments reprinted from various periodicals; critiques by Dr. Carl Behrens of the U.S. Congressional Research Service, Dr. Ralph Lapp, and Dr. Wade Blackman of ERDA; and a response to Dr. Blackman. Responses to Drs. Behrens and Lapp and an exchange with the Council on Energy Independence (Chicago) will appear in a supplementary volume later in 1977. Further exchanges are to be published in *Foreign Affairs:* with Professor Hans Bethe in April 1977 and with Dr. Bertram Wolfe of General Electric in July 1977. The U.S. Atomic Industrial Forum is said to be preparing a critique, but at this writing it is not yet available. A further debate is scheduled for the 14 June 1977 meeting of the American Public Power Association in Toronto, and *Energy Daily* is organizing a June 1977 round-table discussion. The present work and others will be considered in an August 1977 workshop at the Aspen [Colorado] Institute for Humanistic Studies.

It seems a fair summary of the critiques received so far that no errors of fact or logic have been pointed out—a tribute to the many people of diverse views who, in earlier seminars and private discussions, helped so much to sharpen, strengthen, and clarify the arguments.

Meanwhile, as the U.S. debate continues, soft path studies, or elements of analysis that are akin to them and could significantly contribute to them, are underway in many countries. A brief survey of known world activity along these lines as of late February 1977 seems in order.

First, in the United States, the main centers of national research generally along the lines considered in this book are the Center for Environmental Studies, Princeton University (Professor Robert Williams); the Energy and Resources Group, University of California at Berkeley (Professor John Holdren), which is beginning a soft path study for California in cooperation with many other people and groups; the Institute for Environmental Studies, University of

Wisconsin at Madison (Professor John Steinhart); the Union of Concerned Scientists, Cambridge, Massachusetts (Professor Henry Kendall); and, peripherally, the Institute for Energy Analysis, Oak Ridge (Dr. Alvin Weinberg) and the National Energy Strategies Forum, Resources for the Future, Inc., Washington, D.C. State soft path studies are underway in Montana and North Carolina, and are being started in several other states.

Strong Canadian efforts are underway at the Science Council of Canada, Ottawa (Dr. Ray Jackson); the Ministry of Energy, Mines, & Resources, Ottawa (Dr. Peter Dyne); Energy Probe, University of Toronto (Dr. David Brooks); the Canadian Coalition for Nuclear Responsibility, Montréal (Dr. Gordon Edwards); and the Institute for Man and Resources, Charlottetown, Prince Edward Island (Mr. Andy Wells). The last was established by the provincial legislature to study and implement conserver society concepts, and Prince Edward Island is already officially on a soft energy path—a shift now engaging much attention elsewhere in the maritime provinces and being viewed increasingly in Ottawa as a federal demonstration project. Peripheral efforts are also proceeding at centers across Canada, including Petro-Canada, Calgary, and the provincial government of Québec.

In Britain the main centers of soft path research are the Energy Research Group of The Open University, Milton Keynes, Buckinghamshire (Dr. Peter Chapman); the International Institute for Environment & Development, London (Mr. Gerald Leach); and Friends of the Earth Ltd., London (Mr. Walter Patterson). The Science Policy Research Unit at the University of Sussex (Dr. John Surrey) is exploring semisoft energy paths on a world scale. Peripheral efforts, again, exist in Edinburgh, Glasgow, Cambridge, and elsewhere.

In France the established group nearest to soft path work is that of Dr. Jean-Marie Martin, Faculté des Sciences Economiques et Juridiques, Université de Grenoble. The French government has asked the author to set up a national soft path study. The work of the Centre International des Recherches sur l'Environnement et le Développement, Paris (Professor Ignacy Sachs) is also relevant.

In the Netherlands a dispersed but effective group is starting work; basic contacts include Dr. Kees Daey Ouwens, Rijksuniversiteit Utrecht, and Dr. Eric-Jan Tuininga, TNO, Apeldoorn.

In Denmark, pioneering studies of a solar-and-wind-based economy by Dr. Bent Sørensen of Niels Bohr Institutet, København, since extended by a larger group (see Chapter One, note 25), are complemented by the IFIAS project (Dr. Sven Bjørnholm) which the author helped to establish in 1973. Professor Niels Meyer, a physicist at the

Technical Unviersity of Denmark, Lyngby, is studying nonnuclear futures for the EEC and is active in the national studies too.

The Future Studies Group, Stockholm (Dr. Måns Lönnroth) is interested in alternative energy policies, as is the Råd for Natur- og Miljøfag at Universitetet i Oslo (Dr. Paul Hofseth). There is interest in soft paths, though no coherent research program, at the International Institute for Applied Systems Analysis, Laxenburg, Austria (Dr. Wolf Hafele), and private soft path studies may soon begin in Austria. Soft path studies are beginning in Israel under Dr. Tullio Sonnino, Weizmann Institute, Rehovot; and in Italy under the auspices of Dr. Riccardo Galli, Montedison, Milano, with interest also in Ente Nazionale Idrocarburi, Roma (Dr. Marcello Colitti). The Maiden Committee, University of Auckland, New Zealand, has published a 1976 report containing some groundwork for a soft path study, elements of which are being pursued privately, and several analysts affiliated with Friends of the Earth in Australia intend to begin such studies there.

Though no developing country appears to be formally pursuing a soft path study in the sense of this book, several have excellent independent efforts that amount to the same thing. The best appear to be the work of Dr. Amulya K.N. Reddy and colleagues at the Indian Institute of Science, Bangalore, India; of other Indian researchers such as Arjun Makhijani and Jyoti Parikh; and of the alternative technology groups that are widespread in East Asia and the Pacific. Appropriate technology and ecodevelopment are written into the Constitution of Papua New Guinea, whose practical work is particularly impressive. Policies being pursued in Tanzania and the People's Republic of China also have technical affinities to soft energy paths.

As a companion piece to this book, the author is preparing, for FOE/Ballinger copublication later in 1977, an anthology entitled *Energy in Context.* It is to include case studies and summaries of ongoing work from many of the countries listed above, together with contextual essays on biophysical constraints, the relationship of energy policy to macroeconomic and development problems, resilience and how to design for it, the geopolitics of denuclearization, and social transformations. *Energy in Context* will explore further some of the interconnections between energy policy and other concerns, will make concrete and quantitative some concepts treated qualitatively in this book, and will illustrate some of the rich diversity of research now underway on soft energy paths around the world. If it stimulates more people and countries to try to adapt soft path concepts to their own circumstances and to share the results with others, it will have succeeded.

❋

Glossary

α/c: For a selective solar absorber, the ratio of absorptivity at visible wavelengths to emissivity in the medium and far infrared. This ratio is called selectivity, and measures the ability of a well-insulated collector to attain high working temperatures.

ACDA: U.S. Arms Control and Disarmament Agency, State Department.

ACES: Annual Cycle Energy System, a form of energy-conserving building (developed chiefly at Oak Ridge) that pumps heat back and forth between the building and a freezable watertank.

ASME: American Society of Mechanical Engineers.

bar: A unit of pressure almost exactly equal to one normal atmosphere; about 1.02×10^4 kg/m^2 or 14.5 lb/in^2.

baseload: The part of demand on an electrical grid that is constant in time. It is supplied by the power stations of lowest running costs, running as nearly continuously as possible so as not to tie up capital idly.

bbl: Barrel, a liquid measure (usually of oil) equivalent to forty-two U.S. gallons, with a typical energy content of 5.8×10^9 J (5.8 GJ) or 5.5×10^6 BTU. One T of oil is roughly 7 bbl.

bbl/d: Barrel per day, a measure of energy-producing capacity equivalent to 67.1 kW(t).

BCL: Battelle Columbus Laboratory, Columbus, Ohio.

billion: In this book, 10^9 (1,000,000,000), equivalent to the "G" prefix.

BNL: Brookhaven National Laboratory, Upton, Long Island, New York.

BTU: British Thermal Unit, an obsolete unit of energy equivalent to about 1055 J, 2.93×10^{-4} kW-h(t), or 252 gcal.

busbar cost: Cost of producing electricity at the output terminals of a generator or power station (where the bulk electricity is carried by heavy "busbar" conductors), as opposed to cost of electricity delivered to consumers.

°C: Degrees Celsius or Centigrade.

C$: Candian dollars (approximately equal to U.S. dollars).

CdS: Cadmium sulfide, a material that can convert light to electricity by the photovoltaic process. Cadmium is a highly toxic metal.

¢: Cent, 0.01 of a U.S. dollar. (It has a different meaning, irrelevant here, in reactor engineering, as does the dollar.)

CEQ: President's Council on Environmental Quality, Washington, D.C.

CIB: Conseil International pour la Recherche, l'Étude et la Documentation du Bâtiment.

CO_2 : Carbon dioxide, a gas released by burning organic materials.

Con Ed: Consolidated Edison Company, a large New York utility.

constant dollars: Dollars of constant purchasing power, e.g., dollars of the third quarter of 1975.

COP: Coefficient of performance, the ratio of the useful heating or cooling performed by a heat pump (including an air conditioner or refrigerator) to the energy required to operate it. Usually expressed as a dimensionless ratio of kW-h(t)/kW-h(e).

CRS: Congressional Research Service, U.S. Library of Congress.

current dollars: Dollars that have not been reduced to a common basis of constant purchasing power but instead reflect, e.g., anticipated future inflation. They cannot be used for computations unless the assumed inflation rate is stated.

d: Deci-, a prefix equivalent to 0.1.

DC: Direct (as opposed to alternating) current.

D.C.: District of Columbia.

ΔT: Difference of temperature.

dk: Deka-, a prefix equivalent to 10.

(e): A suffix meaning "in the form of electricity."

EEC: European Economic Community, the "Common Market."

EMR: Federal Ministry of Energy, Mines, and Resources, Ottawa.

enthalpy: Loosely speaking, a measure of the quantity of energy without regard to its quality. In terms of enthalpy there is more energy in the form of low temperature heat in the Atlantic Ocean than in the form of oil in the Persian Gulf, but it is of such low quality (high entropy) that it cannot be used to do difficult kinds of work as the oil could.

entropy: A thermodynamic measure of disorder or unavailability. Loosely speaking, a low entropy system is orderly, an unlikely situation brought about by the expenditure of energy; reducing entropy is equivalent to adding information and structure. Increasing entropy, which according to the Second Law of Thermodynamics is the inevitable tendency of all systems, substitutes disorder (ultimately chaos) for order. Any use or conversion of energy irreversibly degrades its quality—increases its entropy—and thus makes it less able to do difficult kinds of work. Electricity and mechanical work (effectively at infinite temperature) are the highest quality forms of energy; then high temperature heat; then low temperature heat, which can do only limited (though important) kinds of work. This explanation will be wholly unsatisfactory to a thermodynamicist, who has a rich theoretical basis for studying entropy, but is the best we can do in this glossary.

EPA: U.S. Environmental Protection Agency.

EPRI: Electric Power Research Institute, Palo Alto, California.

ERDA: U.S. Energy Research and Development Administration (not to be confused with the New York State body NYERDA), created in January 1975 by fissioning the USAEC (the other fission product was the U.S. Nuclear Regulatory Commission or NRC).

°F: Degrees Fahrenheit; $9 \times °C/5 = °F - 32$.

FEA: U.S. Federal Energy Administration.

First Law: The First Law of Thermodynamics, which says that energy (measured by its quantity or enthalpy) is never created or destroyed, but only changed in form and quality. First Law efficiency measures the fraction of the energy supplied to a device or process that it delivers in its output (having no regard to any degradation of energy quality). Thus a furnace with a First Law efficiency of 70 percent delivers 70 percent of the energy present in its fuel as useful heat to the building, while the other 30 percent escapes up the flue or to other useless places.

FOB: Free on board; "FOB New York" means "if delivered at New York, but delivering anywhere else will cost extra."

FOE: Friends of the Earth, the name shared by an international network of independent nonprofit conservation lobbying groups.

FPC: U.S. Federal Power Commission.

F.R.G.: Federal Republic of Germany.

ft: Foot; 12 in = 1 ft = 0.305 m.

G: Giga-, a prefix equivalent to one billion (10^9); thus G$1 is one billion dollars, 1 GJ is one billion joules, etc.

GAO: U.S. General Accounting Office, the auditing arm of Congress.

gcal: Gram-calorie, a unit of energy equivalent to 4.186 J.

GESMO: Generic Environmental Statement on the use of Mixed Oxide fuels in light water reactors (i.e., on plutonium recycle).

GNP: Gross national product.

GPO: U.S. Government Printing Office, Washington, D.C.

GW: Gigawatt (see G), equivalent to 1000 MW or 1 million kW. A large power station has a capacity of order 1 GW(e).

h: Hour.

HMSO: Her Majesty's Stationery Office, London, England.

HUD: U.S. Department of Housing and Urban Development.

IAEA: International Atomic Energy Agency, Vienna.

IEA: As used here, Institute for Energy Analysis, Oak Ridge, Tennessee; also the International Energy Agency, Paris.

IIASA: International Institute for Applied Systems Analysis, Laxenburg, Austria.

in: Inch; 1 in = 1/12 ft = 0.0254 m.

ITC: InterTechnology Corporation, Warrenton, Virginia.

J: Joule, the international unit of energy (see W, gcal, kW, etc.).

k: Kilo-, a prefix meaning one thousand (1,000). Thus k$ is a thousand dollars, kg is a thousand grams, km is a thousand meters, and klb/h is a thousand pounds per hour (a measure of steam production).

°K: Degrees Kelvin or Absolute; $°K = °C + 273.16$.

kV: Kilovolt (thousand volts), a measure of electric potential.

kW: Kilowatt (thousand watts), a measure of energy per unit time. One kW sustained for 1 h (i.e., one kW-h or kilowatt-hour) is equivalent to 3.6×10^6 J, 3413 BTU, 8.6×10^5 gcal, or 1.341 horsepower-h.

lb: Pound, a unit of mass equivalent to 0.4536 kg.

LDC: Less-developed country, one which has not yet made the mistakes of the ODCs (overdeveloped countries).

life cycle cost: A measure of what a thing will cost totally, not only to buy but also to operate over its lifespan. Often computed as capital cost plus present value (q.v.) of running costs.

LMFBR: Liquid-metal-cooled fast neutron breeder reactor.

LNG: Liquefied natural gas.

load duration curve: A type of graph used in utility management, showing the fraction of maximum simultaneous demand expected or experienced as a function of the number of hours per year that it occurs. Baseload demand occurs 8766 hours per year; the most severe peak occurs only a few hours per year (normally in the winter in Europe and in the summer in the U.S.). The flatter (more like a baseload) the curve, the less capital the utility must invest in plants that stand idle much of the time.

LWR: Light-water-moderated and -cooled nuclear reactor.

m: Meter, a unit of length equivalent to 39.36 in. One m^2, the corresponding unit of area, is 10.764 ft^2; one m^3, the corresponding unit of volume, is 1.308 cubic yards or 35.3 ft^3. Also, as a prefix, milli-, or 0.001; thus a m$ is 0.1 ¢.

M. Mega-, a prefix equivalent to one million (10^6). In U.S. engineering, M is used to mean 10^3 (where this book uses k) and MM to mean 10^6 (where this book uses M). The present use of M is consistent with worldwide scientific usage, but the unwary reader is cautioned that MMBTU in U.S. engineering literature is likely to mean 10^6 BTU, not 10^{12}.

M$: Million dollars; not to be confused with m$, thousandths of a dollar.

marginal: Economic jargon for "incremental," "associated with a small future increment in a stock or activity." Specifically, the marginal price of coal, or the price of a marginal ton of coal, or the price of a ton of coal at the margin all mean what you pay for the next ton of coal you buy, not the average of what you paid for coal in the past. The long-run marginal price of coal would be what you would pay for an additional ton at some time well into the future, e.g., after the oil and gas era. If marginal costs are lower than [historic] average costs, then costs are falling; if higher, rising. The world energy system—and probably some other systems too—changed generally from the former to the latter regime sometime around 1970.

merit order: A listing of power stations in an electrical system in order of running costs. The cheapest stations are run baseloaded, the dearest only for infrequent peak loads, and the rest in between according to their merit order.

mi: Mile, a unit of length equivalent to 5280 ft or 1609.4 m.

μm: Micrometer, one millionth of a meter.

mill: In U.S. utility parlance, 0.1 ¢, written here as m$1 or 1 m$.

MIT: Massachusetts Institute of Technology.

mixed oxide: A mixture of UO_2 and PuO_2, used as a nuclear fuel.

MJ: Megajoule, a unit of energy equivalent to roughly 35 grams of coal or 24 of oil, 0.278 kW-h(t), or 948 BTU. See J.

MVA: Megavolt-ampere, a measure of electric power; it is related to MW(e) by a power factor, which reflects how far the electrical load departs from a purely resistive one, so altering the phase angle between current and voltage.

MW: Megawatt, 1 million watts or 1 thousand kW, a measure of energy per unit time. (See kW, W.) Not to be confused with mW, milliwatts or thousandths of a watt.

NAS: U.S. National Academy of Sciences, Washington, D.C.

NASA: U.S. National Aeronautics and Space Administration.

N.J.: New Jersey.

NO_x: Nitrogen oxides, an air pollutant.

NPT: Non-Proliferation Treaty.

NRC: U.S. National Research Council, Washington, D.C.

NSF: U.S. National Science Foundation, Washington, D.C.

NTIS: U.S. National Technical Information Service, Department of Commerce, 5285 Port Royal Road, Springfield, VA 22161; the agency that sells technical documents published by the U.S. government and allied agencies.

OECD: Organization for Economic Cooperation and Development, Paris.

OPEC: Organization of Petroleum Exporting Countries.

ORAU: Oak Ridge Associated Universities, Oak Ridge, Tennessee.

ORNL: Oak Ridge National Laboratory, Oak Ridge, Tennessee.

OTA: Office of Technology Assessment, an arm of the U.S. Congress.

PEI: Prince Edward Island, a maritime province of Canada.

present value: A way of expressing a stream, or series, of future costs as a lump sum equivalent to its total value today. Computed by applying to the cost in each future year a multiplier $1/t^{1+i}$, where t is the future date minus the present date and i, expressed as a decimal (e.g., 0.10 for 10 percent/y), is the annual discount rate assumed. The discounted future costs are then added. The theory is that since a dollar today can be invested at interest i, it is worth $(1 + i)$ dollars next year, $(1 + i)^2$ in two years, and so one; and, conversely, that having a dollar next year is equivalent to having only $1/(1 + i)$ dollars now, etc. If a stream of future costs is the same every year, then their cumulative present value over t years, compounding the inverse interest continuously rather than annu-

ally, is the annual cost times $(e^{it} - 1)/(ie^{it})$, where e = 2.71828. . . .
At commonly used discount rates, say 10 percent/y, the present
value of virtually any common future resource, such as a newborn
child, is essentially zero, since a cost or benefit in twenty years is
worth only 15 percent as much as it would be today, and in fifty
years, less than 1 percent.

price: Cost plus rent. (Rent is the extra charge a seller can exact because of ordinary profit, convenient delivery, low sulfur content
of oil, political connections, monopoly, or other factors extraneous
to technical costs of production. U.S. private utilities sell electricity at a price; many public utilities, especially abroad, sell it at a
cost, i.e., with zero rent.)

Pu: Plutonium.

PWR: Pressurized water reactor, a type of light water reactor (LWR).

q: "Quad", or quadrillion (10^{15}) BTU, equivalent to 26 MT oil. U.S.
primary energy use in 1973 was about 75 q or 2.5 TW. Some
authors use Q, not q, for 10^{15} BTU, but strictly speaking 1 Q =
1000 q, an extremely large unit of energy.

R&D: Research and development; recently extended by ERDA to
RD&D, which adds "demonstration."

reserve margin: Extra generating capacity required to be standing by
for addition to electrical grids, with no particular rapidity, in case
demand is underestimated or normal supplies are unavailable
owing to repairs, nuclear refuelling, etc.; normally calculated as 10
to 20 percent of anticipated peak demand.

RFP: Request for proposal, a document in which a U.S. government
agency asks interested parties to propose R&D projects according
to particular specifications.

s: Second.

Second Law: The Second Law of Thermodynamics (see "entropy").
Second Law efficiency is the ratio of the energy required to do a
task in the best theoretically possible way (according to the thermodynamics of reversible processes, which proceed infinitely
slowly) to the energy actually used to do the task. It thus measures the theoretical room for improvement in end use efficiency,
since thermodynamics enables us to calculate how little energy
can be used to do a job regardless of the method, known or unknown, by which it is done.

Si: Silicon, an element whose crystals can be used to make photovoltaic cells more efficient than CdS cells (q.v.).

spinning reserve: Extra generating capacity required to be standing by for rapid addition to electrical grids in case of failures in generating or transmission equipment. Such failures would leave insufficient capacity operating, and thus would risk major system failures as the frequency of the alternating current dropped below acceptable limits of synchronization. As a precaution, reserve capacity capable of coming into operation in a few seconds must be kept spinning in synchrony with the grid, ready to take up the slack for a few minutes until gas turbine peaking plants and, later, dormant steam plants can be started up. Spinning reserve is normally of two types: hydroelectric (especially pumped storage) plants kept spinning on low water flow, ready to open the water valves to full flow, and steam stations restricted to perhaps 80 percent of normal capacity so that the few minutes' spare steam stored in the boilers can be released by fully opening the steam valves. In either case, the "insurance" for the grid incurs both capital and operating costs that are aggravated by the larger spinning reserve needs of large generating plants.

SRI: Stanford Research Institute, Menlo Park, California.

ston: Short ton (the usual U.S. ton), or 2000 pounds, or 0.91 T.

SW: Separative work, a measure of the energy that must be expended to separate atomic isotopes, e.g., in enriching uranium. Its unit is the separative work unit (SWU), also written as kg (kilograms) of separative work; this is a notional unit that does not refer to kg of material enriched.

(t): A suffix meaning "thermal; in the form of heat or combustible fuels."

T: Tera-, a prefix equivalent to one trillion (10^{12}, one million million). In these terms mankind's present use of commercial energy is at the rate of about 8 TW. Also, metric ton (written "tonne" in Europe), a unit of mass equivalent to 1000 kg or 2204.62 lb or 1.102 ston or 0.984 long tons (English tons of 2240 lb).

T&D: Transmission and distribution (in electrical grids).

turnkey: Sold at an all-inclusive price, ready to operate when the buyer turns a notional key.

U: Uranium, refined at mines to the form U_3O_8 (an oxide called "yellowcake") and later to such forms as UF_6 and UO_2.

UKAEA: United Kingdom Atomic Energy Authority.

UN: United Nations.

USAEC: U.S. Atomic Energy Commission. See ERDA.

USDA: U.S. Department of Agriculture.

USHR: U.S. House of Representatives.

V: Volt, a unit of electrical potential.

VA: Volt-ampere (see MVA); also Virginia.

VAC: Volts of alternating current.

W: Watt, a unit of energy per unit time. One watt is one joule per second (1 J/s); conversely, 1 J = 1 W-s. Human metabolism converts chemical energy at a rate of order 100 W.

y: Year; 8766 h or 3.156×10^7 (approximately $\pi \times 10^7$) s.

In this book, contrary to most European conventions, the period is used as the decimal point, and the comma is used to divide large numbers.

✳

About the Author

Amory Bloch Lovins resigned a Junior Research Fellowship of Merton College, Oxford, in 1971 to become British Representative of Friends of the Earth, Inc., a U.S. nonprofit conservation group. A consultant physicist (mainly in the U.S.) since 1965, he now concentrates on energy and resource strategy. His current or recent clients, none of whom is responsible for his views, include the OECD, several UN agencies, the International Federation of Institutes for Advanced Study, the MIT Workshop on Alternative Energy Strategies, the Science Council of Canada, Petro-Canada, USERDA, the U.S. Office of Technology Assessment, Resources for the Future, and other organizations in several countries. He is active in international energy affairs, has testified before parliamentary and congressional committees, has broadcast extensively, and has published five earlier books, several monographs, and numerous technical papers, articles, and reviews. During the 1977–1978 academic year he will take up a short-term visiting post as Regents' Lecturer at the University of California at Berkeley.